Micronutrient Deficiency in Soils and Plants

Authored By

Theocharis Chatzistathis
School of Agriculture, Forestry and Natural Environment
Aristotle University of Thessaloniki
Greece

CONTENTS

CHAPTERS

FOREWORD

In books based on food sciences, and particularly in those related with the topics of human nutrition, it is recommended to consume fruits and vegetables every day, which are the basic axes of the Cretan diet. Why? Fruits and vegetables are rich in vitamins and mineral nutrients, which are necessary for our alimentation. For fruits and vegetables to be rich in vitamins and mineral nutrients, it is necessary for plants to find nutrients in the soil where they are grown and cultivated. Besides, for good health of plants and trees, it is very important for the soils to contain sufficient quantities of micronutrients for their harmonious growth and development, as well as for an abundant production of fruits and vegetables. This eBook in the sciences of agriculture and environment contribute to a better knowledge of the role of micronutrients in plant nutrition.

Which are these micronutrients, how can we extract them from soil in order to determine their concentrations, why does their deficiency in soils cause problems in plant growth and development, why fruits and vegetables' production is closely related to soil nutrient concentrations? Are there solutions to alleviate micronutrient deficiencies, which are the tools used by agronomists and agriculturists to face a micronutrient constraint in soil and significantly increase yield production, as well as to ameliorate the quality of fruits and vegetables? Which is the strategy that could be used to orientate the conventional agriculture to an ecologically viable agriculture for a sustainable development?

The responses to all these questions are available in the excellent eBook, written by Dr. Theocharis Chatzistathis, investigator in the school of Agriculture, Forestry and Natural Environment, Aristotle University of Thessaloniki, Greece. In this eBook of 9 chapters he describes first of all the 'Micronutrient Solubility and Availability in Soils', followed by 'The Role of Micronutrients in Plant Metabolism & Growth and Diagnostic Tools to Assess Micronutrient Deficiencies'. Afterwards, seven necessary and most important micronutrients for plant growth are presented in different chapters in a very didactic way, which offers to the reader the pleasure to discover them, and in parallel to explore the necessary tools to answer different questions, which could be presented in different situations. This is why the micronutrients like Iron, Zinc, Manganese,

Copper, Boron, Mollybdenium and Chlorine are presented based on very important, up to date, bibliography.

For all the reasons mentioned above, I strongly recommend the study of this eBook not solely to undergraduate and postgraduate students, but also to doctors, researchers, as well to the teaching staff of Agricultural and Environmental Sciences, because it will provide them many answers to very important questions in order to better understand the function of an ecosystem, and particularly they will be able to approach the questions of plant physiology with a different perspective.

Dr. Sevastianos Roussos
Directeur de Recherche IRD
IMBE- Aix Marseille Université
France

PREFACE

Despite the fact that micronutrients are found in very low concentrations in plants, their importance for crop production and plant metabolism is very high and their supply should not be neglected in an integrated fertilization program. They are as essential for plant cycle as macronutrients and the inadequate levels of even one of them may seriously disturb plant metabolism; in that case, characteristic or noncharacteristic micronutrient deficiency symptoms may appear. The lack of micronutrients may disturb photosynthesis, auxin levels, N, carbohydrates and protein metabolism, chlorophyll content, capturing of atmospheric N and many other physiological and metabolic functions of plants.

This eBook aims at presenting and analyzing the problems caused in crop production by micronutrient deficiencies. The necessary plant growth micronutrients, their mobility in soils, their role in plant metabolism, as well as the deficiency symptoms and the critical concentrations are discussed. In addition, the soil factors influencing micronutrient availability are being presented and, finally, some of the most important mechanisms adopted by plants in order to survive under micronutrient deficient soil conditions are included in this eBook, under the light of the most important and recent research data. I really believe that this eBook is a useful tool for MSc and PhD students, as well as for academic personnel and researchers studying mineral nutritional problems in agricultural soils and crops. Furthermore, it may provide with valuable information the growers, giving practical solutions to some topics concerning micronutrient deficiencies and their alleviation, such as fertilization and amendments.

ACKNOWLEDGEMENTS

Declared none.

CONFLICT OF INTEREST

The author confirms that this eBook content has no conflict of interest

Dr. Theocharis Chatzistathis
School of Agriculture, Forestry and Natural Environment
Aristotle University of Thessaloniki
Greece
Tel: +30 2310 998648
E-mail: chchatzi@in.gr

BIOGRAPHY

Dr. Theocharis Chatzistathis has completed his graduate, postgraduate (in soil science) and doctoral (in mineral plant nutrition) studies at the Aristotle University of Thessaloniki, Greece. After receiving his PhD in 2008, he has been educated for a short period of time in sampling and laboratory techniques used for the determination of mycorrhiza, at the Université de Provence, Marseille, France. From 2012 he has been working as scientific cooperator of the department of Horticulture, school of Agriculture, Forestry and Natural Environment of the Aristotle University of Thessaloniki. His current research interests are mainly focused on olive tree's mineral nutrition, plant-soil relationships, as well as on the influence of arbuscular mycorrhiza on plant nutrition. He has published 9 research papers in peer-reviewed SCI journals, 5 book chapters in international published volumes, and many articles in the proceedings of international and national conferences. He has been selected as member of the editorial board of Journal of Agricultural Science and Technology (JAST), David Publishing, and as reviewer from important SCI journals, such as Scientia Horticulturae, Chemosphere, Mycorrhiza, Journal of Plant Nutrition and Soil Science, Acta Physiologiae Plantarum, HortScience and many others, SCI and non-SCI, journals. Apart from Greek (mother language), he also speaks 4 foreign languages: English and French fluently, and Italian and Spanish at a good level.

INTRODUCTION

This eBook has 9 chapters. In chapter 1, the terms 'micronutrient solubility' and 'micronutrient availability' are analyzed, in order the reader to understand micronutrient mobility in soils and their uptake by plants. Furthermore, since micronutrient mobility is influenced by many soil (such as pH, organic matter, $CaCO_3$, cation exchange capacity, soil moisture), plant factors (such as root exudation, development of mycorrhiza in the environment of mycorrhiza), as well as by factors influenced by some human activities (such as crop management practices, over-fertilization with N and P, the choice of the kind of N fertilization used-*i.e.* NO_3^- or NH_4^+), the effect of each of them on micronutrient availability is thoroughly discussed, based on some of the most important research findings published. In the second part of the same chapter, the term 'plant available micronutrient quantities' is analyzed and discussed; in addition, they are presented the different extractants (as well as their ability to determine plant available micronutrient concentrations in soils) that are used by researchers in order to determine micronutrient available quantities. In the second chapter, the role each of the necessary micronutrients has for plant metabolism and growth is analyzed, while some important diagnostic criteria that can be used in order to detect micronutrient starvations and imbalances are also presented and discussed.

Chapters 3-9 deal with specific micronutrient deficiencies (those of Fe, Zn, Mn, Cu, B, Mo and Cl, respectively) and analyze their movement and transformations in soils. The factors influencing micronutrient mobility and availability in soils are fully analyzed and discussed. In addition, the uptake mechanisms, the physiological roles, the symptoms of starvation, the critical concentrations, some of the most important mechanisms of tolerance developed by plants to overcome micronutrient deficiencies, as well as some sources of micronutrients that can be applied either in foliar, or in soil, are included in these chapters. In order to avoid the monotonous repetition of text pages, some important information are presented in Tables and Figures, in the different chapters. In addition, some very characteristic photographs showing micronutrient deficiency symptoms, in order to help the reader obtain a better image of the mineral nutritional imbalances described in text, are included in all chapters.

CHAPTER 1

Micronutrient Solubility and Availability in Soils

Abstract: By the term 'micronutrient availability' we mean the total micronutrient forms in soils, which are available for plants (*i.e.*, all the soluble forms that can be taken up by plants). Availability depends on solubility in soils, *i.e.*, the solubility of micronutrients in soils determines their availability for plants (thus their uptake) and their downward mobility. Understanding the factors controlling trace element solubility allows the selection of soil amendments that promote or reduce their availability and of course the selection of the suitable plant species (those that are optimal for the desired goal of managing trace element influx in the soil-plant system). There are many soil factors influencing micronutrient solubility and availability for plants. The most important of these factors are pH, organic matter content, $CaCO_3$, soil texture, cation exchange capacity (C.E.C.), erosion, soil moisture and temperature *etc*. Generally, trace element solubility in soils depends on a plethora of physical (*e.g.*, water retention capacity), chemical (*e.g.*, pH), microbial (*e.g.*, mycorrhiza) and plant factors, as well as on the properties of each nutrient, which are fully analyzed below. There are many extractants (like DTPA, EDTA, Mehlich-1, Mehlich-3) used to determine available for plants quantities of micronutrients in soils; since great differences exist in the extractability of these chemical solutions, in order to estimate plant available micronutrient concentrations it is absolutely necessary to study micronutrient extractable concentrations in soils, in relation to their uptake by plants. All the topics concerning micronutrient solubility in soils, together with the factors influencing micronutrient availability and uptake by plants, are analyzed in detail in this chapter. In addition, a comparison between the extractant solutions used to estimate plant available quantities of micronutrients in soils is included in the second part of this chapter.

Keywords: $CaCO_3$, cation exchange capacity (C.E.C.), micronutrient availability, micronutrient solubility, organic matter, pH, soil moisture, soil texture, trace elements.

INTRODUCTION

'Micronutrient availability' (all the soluble trace element forms that can be taken up by plants) is closely related to solubility (trace elements can not be available for plants if they are not in soluble forms, *i.e.*, in forms which can be taken up by plants). Generally, fate and bioavailability of metals and trace elements in soils are controlled by three main processes: i) removal of metals from the soil solution by sorption onto soil particles, ii) release of the metal from the soil particle to the soil solution (desorption), iii) precipitation-dissolution of the metal as an

Theocharis Chatzistathis

independent phase in the soil matrix [1]. There are many factors influencing trace element/metal solubility; the most important of these factors are pH, organic matter, $CaCO_3$, cation exchange capacity (C.E.C.), soil texture, soil moisture and temperature, root exudation *etc.* For example, Fe and Mn solubility is positively correlated to soil acidity. Adsorption of some metals (*e.g.,* Zn) onto organic matter controls their solubility and availability for plants [2]; however, in some cases (calcareous or alkaline soils) organometallic complexes (very stable complexes of different metals with organic matter) constitute about 85-90% of the total soluble forms, their solubility is increased with increasing pH (as for example happens in the case of Mn) [3], and they may considerably contribute to micronutrient mineral nutrition. The uptake of these complexes is enhanced by mycorrhiza [4]. From the other factors influencing trace element availability, they should be distinguished: i) the high bicarbonate content, which decreases trace element solubility and uptake by plants (Fe deficiency and chlorosis of plants is the most characteristic example) and ii) soil moisture, which influences trace element solubility *via* the prevalence of soluble/non soluble micronutrient forms under aerobic or anaerobic conditions, respectively. Finally, soils with high C.E.C. have the ability to exchange micronutrient cations between soil solution and the negatively charged surfaces of clay and organic matter, so they protect trace elements from leaching.

Micronutrient concentrations and availability greatly varies among soil types; responsible for this variation is the chemical composition of the parent material from which soil types were originated from. For example, high Ni concentration in ultramafic (serpentine) soils is a characteristic example of the influence of parent material on metal availability. Furthermore, Mn concentration of different rocks varies a lot; the greatest Mn contents are found in basic-eruptive rocks (basalt, gabbro), varying between 1000 and 2000 p.p.m. [5]. According to Robinson *et al.* (2009), trace element loading in soils is a function of parent material plus subsequent atmospheric or waterborne deposition [6]. For example, Mn concentration in soils varies from traces (in some Podzol, highly leached, soils of Poland) up to more than 10000 p.p.m. (in unleached alkali soils of Chad) [5].

The purpose of this chapter is to collect, present and discuss all the scientific information concerning micronutrient availability in soils and the factors (soil, plant and other ones, affected by human activities) by which it is influenced. Furthermore, particular emphasis is given to the extractant solutions that are used in order to estimate micronutrient plant available concentrations in soils, under the light of the most important and recent research and review articles.

Factors Influencing Micronutrient Availability

Micronutrient availability in soils greatly varies, depending on parent material, soil properties (pH, organic matter, $CaCO_3$, cation exchange capacity, soil moisture and temperature *etc.*) and climatic conditions. Most of the trace elements may be found in low concentrations (deficient or sufficient for plant nutrition) in soils, while in other cases, where very high concentrations occur, they are considered to be heavy metal contaminants. A better understanding of the mechanisms that control free ion (Fe^{2+}, Mn^{2+}, Zn^{2+}, Cu^{2+}, Cd^{2+} *etc.*) activity could contribute to the choice of the correct soil amendments/treatments in order to avoid micronutrient deficient or metal toxic conditions for plants. For example, possible controlling mechanisms for Zn^{2+} activity include adsorption or precipitation. According to Catlett *et al.* (2002), soil organic carbon and pH were statistically significant parameters in a multiple regression with log Zn^{2+} activity. The significance of the organic carbon might suggest that the adsorption onto organic matter controlled Zn solubility in some of the soils they studied [2]. Generally, for some micronutrients (especially for Fe and Mn) the correlation between soil acidity and ion activity is positive, *i.e.*, as soil acidity increases ion activity (Fe^{2+} and Mn^{2+}) also increases. For other micronutrients, such as B and Mo, the optimum pH range for maximum availability varies from 5.0 to 7.0 and from 9.0 to 10.0 for B, and from 7.0 to 10.0 for Mo, respectively. Apart from pH and organic carbon, $CaCO_3$ is another crucial factor influencing micronutrient availability in soils. Under high free $CaCO_3$ conditions micronutrient availability significantly decreases. In Mediterranean soils, where high pH and $CaCO_3$ content together with low organic matter is often a reality, micronutrient constraints are many times serious mineral nutritional problems [7]. Generally, the most frequent occurring deficiencies are related to extremely acid light sandy soils, or to alkaline (calcareous) soils with improper water regimes and with excesses of phosphate, N and Ca, as well as with Fe and Mn oxides [3]. All these factors influencing micronutrient availability in soils can be divided into two categories (soil and non-soil) and they are fully analyzed and discussed below.

Soil Factors

I. **pH.** pH is the most important factor influencing micronutrient availability in soils. In strongly alkaline soils, high pH favors the formation of insoluble micronutrient forms, *i.e.*, those which can not be absorbed by plants. Generally, as soil pH increases, the availability of most micronutrients decreases, with the exception of Mo [2]. As for example Mn, Fe and Zn availability is maximum in pH range from 5 to 6.5 and minimum in alkaline soils, Mo deficiency is not a problem in Mediterranean-type soils (which are usually alkaline), since its availability is high in alkaline soils [7]; Mo deficiency is a nutritional problem in acid soils [4].

II. **Organic matter.** Soil organic matter has been of particular interest in studies of heavy metal sorption by soils, because of its significant impact on cation exchange capacity (C.E.C.), and more importantly, on the tendency of transition-metal cations to form stable complexes with organic ligands [9]. The formation of these stable and strong complexes explains why rich in organic matter soils are more susceptible in micronutrient deficiency, than the poor ones (the negatively charged surfaces of organic macromolecules adsorb micronutrient cations which usually happens with Cu^{2+}- and form strong organometallic complexes, which can not be absorbed by plants) [4]. The amount of soil organic matter affects the binding of heavy metals and speciation on soil solution [10]. Adsorption onto organic matter controlled Zn availability in some of the soils studied by Catlett *et al.* (2002) [2]. However, in calcareous and alkaline soils (where micronutrient deficiency is often a serious problem) organometallic complexes may significantly contribute to the improvement of micronutrient mineral nutrition of higher plants. The solubility of these complexes increase with increasing pH (as for example happens in the case of Mn; [3]) and their role in the contribution of micronutrient mineral nutrition of higher plants is often valuable. This is the reason why under calcareous and alkaline soil conditions Mn and Zn constraints are not necessarily nutritional problem for plants [11]. According to McBride *et al.* (1997) [12], the organic matter content is often, but not always, a statistically significant variable parameter in predicting metal solubility from soil properties. It seems that for Cu at least, solid organic matter limits free metal activity, whilst dissolved organic matter promotes metal solubility in well-aged soils with respect to the metal pollutant [12]. According to Kasmaei and Fekri (2012) [13], who studied the effect of organic matter on the release, behaviour and extractability of Cu and Cd in two soils of Iran, the kind of organic matter plays a very important role in metal extractability and solubility. More specifically, extractability of Cu increased with poultry manure treatment and decreased with pistachio compost treatment, as compared to the control soil, while extractability of Cd decreased with both pistachio compost and poultry manure, compared to the control soil [13].

III.   **$CaCO_3$.** Increased $CaCO_3$ content generally decreases the solubility and uptake of most micronutrients. Iron deficiency is the most common micronutrient constraint provoked by high bicarbonate content. Under such soil conditions Fe deficiency symptoms for crops, such as leaf chlorosis, occur.

IV.   **Cation exchange capacity (C.E.C.).** The colloid surfaces of clay are negatively charged, so they have the ability to adsorb and exchange cations, thus to protect them from leaching. The ability of the solid phase to exchange cations, the so-called C.E.C., is one of the most important soil properties governing the cycling of trace elements in soil. The excess amount of adsorbed cations, compared to the amount in solution, is interpreted as the buffering capacity of soils, while adsorption capacity defines the amount of ions needed to occupy all adsorption sites per unit of mass. The C.E.C. of different soils varies widely, both in quantity and quality, and can range from 1 to 100 meq/100g soil (however, C.E.C. of most soils does not exceed the value of 30) [3]. High C.E.C. is a characteristic of soils with high clay and organic matter content. Organic matter significantly contributes to the C.E.C. of many soils [14]. In light textured soils, where low clay and organic matter contents occur, micronutrient deficiencies may be a reality due to enhanced leaching from the upper to the deeper soil layers.

V.   **Soil texture.** In leached acid sandy soils, where micronutrient cations are leached from the upper to the deeper soil layers (as happens with macronutrient cations), micronutrient availability is significantly reduced [4]. In contrast to that, soils with high clay content (fine texture) are less likely to be low in plant available micronutrients [8].

VI.   **Soil moisture and temperature.** Soil moisture and temperature affect micronutrient availability and uptake by plants. The shortage of soil moisture negatively affects micronutrient solubility. On the other hand, excess soil moisture may lead to the prevalence of anaerobic and reducing conditions. Under these soil conditions the solubility of some micronutrients (such as that of Mn) increases (convert of Mn^{4+}, unavailable for plants form, to Mn^{2+}, which is the soluble, available for plants, form), while that of others, such as Zn, decreases (conversion of Zn^{2+} to Zn^{+}, which is the form that can not be absorbed by plants). Another negative aspect of excess soil moisture is the enhanced leaching of micronutrients from the upper to the deeper layers, especially in light textured soils. Finally, low soil temperatures negatively affect micronutrient uptake by plants, especially when they are combined with wet soil environments [4]; in that case, during cold seasons plants may have low levels of some trace elements, thus micronutrient deficiency may be caused in grazing animals [3]. Miyasaka and Grunes (1997) [15] found that increasing soil temperature from 8°C to 16°C increased by twofold the contents of Cu, Zn and Mn in leaves of winter wheat.

Factors Influenced by Human Activities

I.   **The intensification of cultivation, and the crop management practices adopted by the farmers.** In the intensified cultivated soils great quantities of micronutrients are came away from the orchards due to harvesting and pruning [4]. In contrast to that, Wei *et al.* (2006) [16] found that long term cropping and fertilization altered several important soil properties and increased the plant available micronutrient content. Between manure and synthetic fertilizers, the first one seems to be a better source of available Mn, Fe and Zn, compared to the second [17]. According to our opinion, the differences between the findings of different authors concerning micronutrient availability under crop cultivation are probably owed to the kind of management practices adopted by the farmers (hyper-intensified systems of cultivation with high inputs of fertilizers and irrigation water, or less intensified ones with the minimal, absolutely necessary use of fertilizers, water, insecticides and fungicides, as well as with the minimum use of heavy agricultural machines).

II.   **Soil erosion.** In the areas where soil erosion happens due to human activities, great quantities of micronutrients from the upper soil layers are lost. Micronutrients in the upper soil layers may be found in exchangeable, or organically bound form [4].

III. **Over-fertilization with N and P fertilizers.** In P over-fertilized soils, some insoluble phosphate substances of trace elements (especially Mn, Fe and Zn phosphate substances) may be formed, so micronutrient solubility is considerably reduced and trace elements are not available to plants [4]. High P rates induced Cu and Zn deficiency in hybrid *Poplar* sp. trees, although DTPA-extractable micronutrient status in soil was not affected except for Mn; these deficiencies were probably ascribed to leaf P/Zn and P/Cu ratios among clones [18]. So, induced micronutrient deficiency is one of the basic reasons why we should do modest and not excessive P fertilizations. According to Habib (2012), the imbalanced use of chemical fertilizers, especially that of nitrogen and phosphorus, may reduce seed quality and cause micronutrient deficiency in crops [19]. In contrast to these findings concerning N fertilizers, Losak *et al.* (2011) [20] found that the continuous single use of N fertilization has not so far resulted in a micronutrient deficiency of maize (*Zea mays* L.) plants, limiting the nutrient density of the grain or reducing its quality. Furthermore, Kutman *et al.* (2011) [21] found that high N elevated the endosperm Fe concentration of wheat plants up to 100%. According to the same authors, improved N nutrition is effective in increasing Zn and Fe of the whole grain and particularly the endosperm fraction, and might be a promising strategy for tackling micronutrient deficiencies in countries where white flour is extensively consumed. However, even the very high N level used in the study of Kutman *et al.* (2011) was not excessive, according to their remark [21]. The interaction between N and micronutrients is not well understood; this is why reports in the literature (basically concerning Zn) are often inconsistent [20]. In our opinion, the inconsistencies in the literature concerning the interaction between N and micronutrients may be probably ascribed to the level of N fertilization applied to soil (excessive or not), as well as to the period (prolonged, for many years, or short) of the use of N fertilizers in crops. The most possible explanation for micronutrient deficiency induced by N over-fertilization is that increased levels of N may increase the requirement for micronutrients due to increased growth (dilution effect).

IV.   **The kind of N-fertilization.** The choice of the kind of N-fertilization used for plants plays a crucial role in the acidification of the rhizosphere environment. More specifically, when NH_4^+ fertilization is preferred (during the winter months), rhizosphere is acidified (due to the exudations of H^+ from the root system of plants), so the availability of many trace elements (Mn, Fe, Zn) is increased. Tong *et al.* (1997) [22] found increased solubilisation of Mn in the rhizosphere of barley plants, fed with NH_4^+. In contrast to that, when NO_3^- fertilization is chosen (during the vegetative period during spring and summer) OH^- are exudated from the roots, pH of the rhizosphere is increased, and micronutrient solubility is usually decreased, as found for Mn [23].

Plant Factors

I.   **Root exudates.** Some plant species have the ability to produce organic exudates (organic substances of low molecular weight produced during the period of maximum growth rate of plants, which favor micronutrient solubility and uptake through the form of organic complexes) from their root system in order to enhance micronutrient solubility [24, 25]. According to Marschner *et al.* (1986) [24], in response to Fe deficiency, roots of grass species release non-proteinogenic amino-acids ('phytosiderophores'), which dissolve inorganic Fe compounds by chelation of Fe(III) and also mediate the plasma membrane transport of this chelated Fe into the roots. More specifically, in the root system of *Lupinus albus* L., soluble Fe phosphates are mobilized by the exudation of chelating substances (probably citrate), net excretion of H^+ and increase in the reducing capacity [20]. Mugineic acid is another typical example of this kind of exudates and is responsible, according to Maruyama *et al.* (2005) [26], for the differential Fe acquisition ability between barley and rice. Furthermore, Chatzistathis *et al.* (2009) refer that a similar mechanism could be probably responsible for the differential Mn and Fe uptake efficiency and accumulation between the olive cultivars 'Koroneiki' and 'Kothreiki' [27]. Finally, differences between genotypes concerning their ability to tolerate Zn deficiency may be ascribed, according to Rengel (1997), to greater released amounts of the phytosiderophore 2-deoxymugineic acid by tolerant genotypes, than by sensitive ones [28].

II. **Mycorrhiza.** Mycorrhiza is a symbiosis between plant and mycorrhizal fungus, which is mutually beneficial for both parts. The benefit for plant is the enhanced nutrient uptake (especially under poor nutrient conditions), while for fungus is the supply of photosynthetic products, provided by photosynthetic operation. Mycorrhiza has the ability to favor nutrient solubility and uptake, especially under alkaline conditions, where micronutrient deficiency (for example Fe and Mn) is a reality. Under these adverse soil conditions, mycorrhiza enhances the uptake of organometallic complexes, which many times constitute about 85% of the total soluble micronutrient forms [4, 11]. Despite the fact that mycorrhiza enhances nutrient uptake, sometimes the capacity of root exudates for micronutrient solubilization may be decreased in mycorrhized plants; according to Posta *et al.* (1994) [29], the root exudations of maize plants inoculated with the fungus *Glomus mosseae* had two times lower capacity for Mn solubilization, than the root exudations produced by non-inoculated maize plants.

Micronutrient Extractants Used to Determine Available Concentrations of Trace Elements in Soils

Many extractant solutions have been used in order to determine micronutrient plant available quantities in soils; generally, they can be divided into four groups: a) distilled water and salt solutions ($CaCl_2$, $MgCl_2$, $NaNO_3$, NH_4NO_3), b) CH_3COONH_4 (pH 4 and 7), c) acid solutions, like 0.1N H_3PO_4 or a combination of 0.1M HCl and 0.033 M H_2SO_4, or Mehlich-3 solution and d) organic extractant solutions, such as EDTA or DTPA [30]. The micronutrient quantities that are extracted by acid extractants are greater than those extracted by other solutions, since they can extract nearly total amounts of trace elements; Indeed, Garcia *et al.* (1997) [31], who compared Mehlich-3, EDTA, DTPA and Soltanpour & Schwab concerning their ability to determine micronutrient availability in 4 Argentinean soils, found that Mehlich-3 yielded the highest extractions for Fe, Mn, Zn and Cu, compared to the other extractants. Furthermore, Chatzistathis *et al.* (2014) found that Mehlich-3 extracted significantly greater quantities of Mn, Fe and Zn, compared to those extracted by DTPA and Soltanpour & Schwab, in three soil types of Macedonia, northern Greece [32]. Cancela *et al.* (2002) also found that in acid soils greater quantities of Cu, Fe, Zn and Mn were extracted by Mehlich-3 solution, than by DTPA [33]. However, between organic extractants similar amounts of micronutrients are usually extracted; Martens and Lindsay (1990) [34] found that DTPA and Soltanpour & Schwab tests extracted about the same amount of Fe, while the second method extracted 0.3, 0.8 and 0.5 mg/kg more Cu, Mn and Zn respectively. In accordance to the results of Martens and Lindsay (1990) [34] are those of Garcia *et al.* (1997) [31], who found that DTPA and Soltanpour & Schwab solutions extracted similar amounts of Zn, Fe and Mn in Argentinean soils. Generally, chelating agents and buffered salt solutions are believed to extract potentially mobile portions of metals, while neutral salt solutions have been introduced as simulating the natural soil solutions and therefore are useful to evaluate the ecological relevance of metals [3].

Coca Cola has been proposed as a relatively simple extractant and is easily applied in soil testing procedures for mobile trace elements. The active compound in Coca Cola is phosphoric acid, thus the overall extraction was similar to that of commonly used phosphoric acid methods. Compared to DTPA, Coca Cola extracted only 27% of Fe, 38% of Cu, 86% of Zn, but 165% of Mn [35]. Due to the relatively stable composition of Coca Cola, such an extractant might be suitable for specific experiments and projects [3]. Usually, most of the micronutrient soil extractants that are used determine greater micronutrient concentrations than the plant available ones. This happens because these solutions determine not only the easily plant available quantities of micronutrients, but also those that are attached to cation exchange sites, as well as those from other soil 'pools' (Cu, Fe, Mn and Zn hydroxides) [36]. In Table **1** is presented the relative phytoavailability of different species of trace elements from soils. As it is clear from that table, easily phytoavailable are only the simple or complex micronutrient cations in soil solution, while moderately phytoavailable are the exchangeable ones in organic and inorganic soil surfaces (Table **1**).

Table 1. Relative phytoavailability of different species of trace elements in soils (from Kabata-Pendias, 2001) [3].

Trace Elements Species	Phytoavailability
Simple or complex micronutrient cations in solution phase	Easy
Exchangeable cations	Medium
Chelated cations	Slight
Micronutrient compounds precipitated on soil particles	Available after dissolution
Micronutrients bound or fixed inside organic substances	Available after decomposition
Micronutrients bound or fixed inside mineral particles	Available after weathering and/or decomposition

So, the crucial matter is which of the extractants used by researchers determine the 'real', available for plants, micronutrient quantities (micronutrient quantities that can be taken up by plants). For that reason, soil micronutrient extractable quantities determined by various extractants should be studied in relation to micronutrient uptake by plants, as well as to the produced biomass and crop yields [37]. Correlation coefficient between micronutrient uptake by plants (or leaf micronutrient concentrations) and micronutrient extractable quantities in soils, determined by various extractants, is usually used by most researchers in order to indicate the most reliable extractant solution (that which best approaches the available micronutrient quantities, *i.e.*, those that can be absorbed by plants).

Tables **2** and **3** show the correlation coefficients between extractable quantities of Mn, Fe and Zn determined by three extractants in three soil types of northern Greece and i) total plant micronutrient content, ii) leaf Mn, Fe and Zn concentrations in olive plants, respectively. As it is clear from these tables, in the case of Mn, the highest correlation coefficients between extractable Mn concentration and leaf Mn, as well as total per plant Mn content, were found when DTPA was chosen as extractant, in all soil types. In the cases of Fe and Zn, soil type played a very important role in the indication of the most reliable extractant solution (that which best correlates plant Mn and Fe to extractable micronutrient quantities). In contrast to these results, Sequeira *et al.* (2011) found in their study that the highest R^2 value between soil extractable Cu, Zn and Mn concentrations and foliar Cu, Zn and Mn contens of *Eucalyptus* trees achieved with Mehlich-1 (thus, it was the most reliable extractant to evaluate micronutrient availability in commercial *Eucalyptus* plantations), than with Mehlich-

Table 2.   Correlation coefficient (R) between soil micronutrient extractable concentrations and leaf Mn, Fe and Zn concentrations of the cultivar 'Chondrolia Chalkidikis' [32].

Soil Type	Micronutrient Extractant Solution		
	DTPA	Mehlich-3	Soltanpour & Schwab
	Mn		
Marl	0.86	N.S.	0.58
Gneiss schist	0.80	N.S.	N.S.
Peridotite	0.88	N.S.	N.S.
	Fe		
Marl	N.S.	0.51	0.44
Gneiss schist	0.94	N.S.	0.47
Peridotite	0.38	0.57	N.S.
	Zn		
Marl	0.51	N.S.	0.27
Gneiss schist	N.S.	N.S.	0.60
Peridotite	0.71	N.S.	N.S.

The correlation coefficients were calculated by the SPSS statistical program (N.S.=Non significant).

3 or DTPA [38]. According to our experience, not only soil type, but also plant species plays a crucial role in the indication of the most reliable extractant solution (that which best evaluates micronutrient availability for crops). This is probably the

reason for the differences often found between researchers in the indication of the most appropriate extractant (like in the case between our results for olive plants and those of Sequeira *et al.* (2011) for *Eucalyptus* plantations).

Table 3. Correlation coefficient (R) between soil micronutrient extractable concentrations and total per plant uptake of Mn, Fe and Zn by the cultivar 'Chondrolia Chalkidikis' [32].

Soil Type	Micronutrient Extractant Solution		
	DTPA	Mehlich-3	Soltanpour & Schwab
	Mn		
Marl	0.82	N.S.	0.48
Gneiss schist	0.76	N.S.	N.S.
Peridotite	0.80	N.S.	N.S.
	Fe		
Marl	0.31	0.55	0.37
Gneiss schist	0.72	N.S.	0.35
Peridotite	0.40	0.62	N.S.
	Zn		
Marl	0.64	N.S.	0.36
Gneiss schist	N.S.	N.S.	0.49
Peridotite	0.78	N.S.	0.38

The correlation coefficients were calculated by the SPSS statistical program (N.S.=Non significant).

CONCLUSION

In conclusion, micronutrient availability (*i.e.* the micronutrient forms that can be absorbed by plants) in soils is closely related to micronutrient solubility. The factors that influence trace element availability can be divided into three groups: a) soil factors (pH, organic matter, C.E.C., $CaCO_3$, soil moisture, soil texture), b) plant factors (root exudation, mycorrhiza) and c) factors influenced by human activities (intensification of cultivated soils, soil erosion, abuse of P fertilizers, the kind of N fertilization-NH_4^+ or NO_3^- chosen by the farmers). Trace element concentrations and availability in soils depend also greatly on chemical composition of parent material, as well as on the climatic conditions of each region.

There are many extractants that are used in order to determine micronutrient plant available concentrations in soils: Organic extractants (DTPA, EDTA *etc.*), acid solutions (HCl, H_2SO_4, H_3PO_4, Mehlich-1, Mehlich-3), CH_3COONH_4 (pH 4 and 7) solutions, and distilled water and salt solutions. The acid solutions extract the highest concentrations of trace elements in soils, compared to other extractants, due to the fact that they determine not only the plant available quantities of micronutrients, but also those that are attached to cation exchange sites, as well as those from other soil 'pools' (like Cu, Fe, Mn and Zn hydroxides). Generally, the most reliable micronutrient extractant is that which best correlates extractable concentrations in soils to total plant micronutrient content, or leaf micronutrient concentrations. For that purpose (in order to indicate the most reliable extractant solution), the correlation coefficient between plant micronutrient concentrations and extractable ones in soils determined by different extractants, is usually used by most researchers. From some of our recent research data it was found that not only the trace element studied (Fe, Mn, Cu, Zn, *etc.*), but also soil type plays a crucial role in the indication of the most reliable extractant solution.

REFERENCES

[1]   Sparks DL. Environmental soil chemistry, 2nd ed. 2003. New York: Academic Press.
[2]   Catlett KM, Heilb DM, Lindsayc WL, Ebingerd MH. Soil chemical properties controlling zinc activity in 18 Colorado soils. Soil Sci Soc Am J 2002; 66(4): 1182-1189.
[3]   Kabata A, Pendias H. Trace elements in soils and plants. 3rd ed. CRC Press, USA, 2001.
[4]   Alifragis D. Soil: Genesis, Properties and Classification. Volume I. Aivazi Publications, Thessaloniki, Greece 2008; pp. 487-492. In Greek.
[5]   Aubert H, Pinta M. Trace elements in soils. Elsevier Scientific Publishing Company, 1977; pp. 43-53.
[6]   Robinson BH, Banuelos G, Conesa HM, Evangelou MW, Schulin R. The phytomanagement of trace elements in soil. Critical Rev Plant Sci 2009; 28: 240-266.
[7]   Rashid A, Ryan J. Micronutrient constraints to crop production in soils with Mediterranean type characteristics: A review. J Plant Nutrition 2004; 27(6): 959-975.
[8]   McKenzie RH. Micronutrient requirements of crops. Alberta: Agriculture and Rural Environment; 1992: http://www1.agric.gov.ab.ca/$department/deptdocs.nsf/all/agdex713.
[9]   Elliott HA, Liberati MR, Huang CP. Competitive adsorption of heavy metals by soils. J Environ Quality 1986; 15: 214-219.
[10]   Lo KSL, Yang WF, Lin YC. Effects of organic matter on the specific adsorption of heavy metals by soil. Toxicol Environ Chemistry 1992; 34: 139-153.
[11]   Keramidas B. Fertility of Soils. Publications of the Aristotle University of Thessaloniki, Thessaloniki, Greece; 1997, pp. 73-74. (In Greek).

[12]   McBride M, Sauve S, Hendershot W. Solubility control of Cu, Zn, Cd and Pb in contaminated soils. Europ J Soil Sci 1997; 48(2): 337-346.

[13]   Kasmaei LS, Fekri M. Effect of organic matter on the release behavior and extractability of Cu and Cd in soil. Com Soil Sci Plant Anal 2012; 43: 2209-2217.

[14]   Hodgson JF. Chemistry of the micronutrient elements in soils. In: Norman AG, Ed. Advances in Agronomy 1963; 15: 119-159.

[15]   Miyasaka SC, Grunes DL. Root zone temperature and calcium effects on phosphorus, sulphur and micronutrients in winter wheat forage. Agron J 1997; 89: 742-748.

[16]   Wei X, Hao M, Shao M, Gale WJ. Change in soil properties and the availability of soil micronutrients after 18 years of cropping and fertilization. Handbook of Environmental Chemistry, Volume 5: Water Pollution 2006; 91: 120-130.

[17]   Gao M, Che FC, Wei CF, Xie DT, Jang JH. Effect of long-term application of manures on forms of Fe, Mn, Cu and Zn in purple paddy soil. Plant Nutr Fertil Sci 2000; 6: 11-17.

[18]   Teng Y, Timmer VR. Phosphorus-induced micronutrient disorders in hybrid poplar. I. Preliminary diagnosis. Plant Soil 1990; 126: 19-29.

[19]   Habib M. Effect of supplementary nutrition with Fe, Zn chelates and urea on wheat quality and quantity. Afr J Biotech 2012; 11: 2661-2665.

[20]   Losak T, Hlusek J, Martinec J, *et al*. Nitrogen fertilization does not affect micronutrient uptake in grain maize (*Zea mays* L.). Acta Agric Scand B Soil Plant Sci 2011; 61: 543-550.

[21]   Kutman UB, Yildiz B, Cakmak I. Improved nitrogen status enhances zinc and iron concentrations both in the whole grain and the endosperm fraction of wheat. J Cereal Sci 2011; 53: 118-125.

[22]   Tong Y, Rengel Z, Graham RD. Interactions between nitrogen and manganese nutrition of barley genotypes differing in manganese efficiency. Ann Bot 1997; 79: 53-58.

[23]   Grass LB, MacKenzie AJ, Meek BD, Spencer WF. Manganese and Iron solubility changes as a factor in tile drain clogging: I. Observations during flooding and drying. Soil Sci Soc Am Proc 1973; 37: 14-17.

[24]   Marschner H, Romheld V, Horst WJ, Martin P. Root induced changes in the rhizosphere: Importance for the mineral nutrition of plants. Z. Pflanzenernaehr. Bodenk 1986; 149: 441-456.

[25]   Smith KA, Paterson JE. Manganese and Cobalt. In 'Heavy metals in soils' 1990; pp. 225-243.

[26]   Maruyama T, Higuchi K, Yoshida M, Tadano T. Comparison of iron availability in leaves of barley and rice. Soil Sci Plant Nutr 2005; 51: 1037-1042.

[27]   Chatzistathis T, Therios I, Alifragis D. Differential uptake, distribution within tissues, and use efficiency of Manganese, Iron and Zinc by olive cultivars 'Kothreiki' and 'Koroneiki'. HortSci 2009; 44: 1994-1999.

[28]   Rengel Z. Root exudation and microflora populations in rhizosphere of crop genotypes differing in tolerance to micronutrient deficiency. Plant Soil 1997; 196: 255-260.

[29]   Posta K, Marschner H, Romheld V. Manganese reduction in the rhizosphere of mycorrhizal and nonmycorrhizal maize. Mycorrhiza 1994; 5(2): 119-124.

[30]   Reisenauer HM. Determination of plant-available soil manganese. In: Graham RD, Hannam RJ, Uren NC, Eds. Manganese in soils and plants, Proceedings of the International symposium on 'Manganese in soils and plants'. Kluwer Academic publishers, 1988; pp. 87-98.

[31]   Garcia A, De Iorio AF, Barros M, Bargiela M, Rendina A. Comparison of soil tests to determine micronutrients status in Argentina soils. Com Soil Sci Plant Anal 1997; 19-20: 1777-1792.

[32]   Chatzistathis T, Alifragis D, Therios I, Dimassi K. Comparison of three micronutrient soil test extractants in three Greek soil types. Com Soil Sci Plant Anal 2014; 45: 381-391.

[33]   Cancela RC, De Abreu CA, Paz-Gonzalez A. DTPA and Mehlich-3 micronutrient extractability in natural soils. Com Soil Sci Plant Anal 2002; 33: 2879-2893.

[34]   Martens DC, Lindsay WL. Testing soils for Copper, Iron, Manganese and Zinc. In: Westerman RL, Ed. Soil Testing and Plant Analysis, 3rd ed. Soil Sci Soc Am J, Madison, WI, USA, 1990; pp. 229-260.

[35]   Schnug E, Flechenstein J, Haneklaus S. Coca Cola is it! The ubiquitous extractant for micronutrients in soil. Com Soil Sci Plant Anal 1996; 27: 1721-1730.

[36]   Mehlich A. Mehlich 3 Soil test extractant: A modification of Mehlich 2 extractant. Com Soil Sci Plant Anal 1984; 15: 1409-1416.

[37]   Black CA. Nutrient supplies and crop yields: Response curves. In: Soil Fertility and Control. CRC Press 1993; pp. 1-5.

[38]   Sequeira CH, Baros NF, Lima-Neves JC, Novais RF, Silva IR, Alley M. Micronutrient soil-test levels and *Eucalyptus* foliar contents. Com Soil Sci Plant Anal 2011; 42: 475-488.

CHAPTER 2

The Role of Micronutrients in Plant Metabolism & Growth and Diagnostic Tools to Assess Micronutrient Deficiencies

Abstract: Despite the fact that micronutrients are required in very low concentrations by plants, they are as essential for plant metabolism and growth as macronutrients. Foliar analysis is a valuable tool in order to detect micronutrient deficiencies before macroscopic symptoms appear in plants; for that purpose, critical micronutrient concentrations have been established. Other diagnostic tools used to assess micronutrient deficiencies are soil analysis, plant-growth response (in annual plants), and visual observation of symptoms. Recently, more and more biochemical indicators, as early detectors of micronutrient deficiencies, are used. Fertilization (soil or foliar application) should be included in the cultivation program in order to improve the low nutritional status of plants. However, before fertilization, it is absolutely necessary to have a deep knowledge of the physiological roles of the necessary micronutrients for normal plant growth. So, the physiological roles of micronutrients, as well as some critical deficiency concentrations in soils and plants, are presented in detail and fully discussed in this chapter.

Keywords: Critical concentrations, foliar deficiency symptoms, micronutrient deficiency, micronutrient sufficiency, plant growth, plant metabolism, physiological roles.

INTRODUCTION

Micronutrients are present at low concentrations (mg/kg d.w. or less) in soils and plants. Cu, Zn, Fe, Mn, Mo, B, and Cl are essential for the normal growth of plants in traces (very low concentrations). However, when trace elements are found in high concentrations they are considered to be heavy metal pollutants [1]; in heavy metals are also included Cd, Pb, Cr, Hg and As *etc.* Some plants can tolerate and accumulate much higher concentrations of trace elements than regular plants. These plants are called 'hyper-accumulators' and they can be used for phytoremediation [2].

In order to decide about the essential fertilization program for plants, it is absolutely necessary to study the critical deficiency concentrations of Fe, Mn, Zn, Cu, Mo and B. Information in detail on critical deficiency concentrations for each one of these trace elements is given in chapters 3-7; however, some critical

Theocharis Chatzistathis

concentrations are also included in this chapter. In order to study the critical micronutrient levels of a crop, leaf nutrient analysis is used; the micronutrient concentrations determined by leaf analysis are compared to those of the optimum ones (for maximum plant growth). So, leaf nutrient analysis is a useful tool to study the nutritional status of plants; however, a deep knowledge of the physiological roles of the 6 indispensable for plant growth micronutrients is also necessary. In the next paragraph, the physiological roles of Cu, Zn, Fe, Mn, Mo and B are fully analyzed.

The purpose of this chapter is to fully analyze the physiological roles of the necessary for plants micronutrients, as well as to present some of the most important symptoms of micronutrient deficiencies and to discuss about critical concentrations for normal plant growth. The diagnostic tools used to assess micronutrient deficiencies in crops are also included in this chapter and discussed the most important advantages and disadvantages of each one of them. Finally, some indicative examples of the input/output budgets of micronutrients in agro-ecosystems, as well as in some natural ecosystems, are also included in this chapter.

Physiological Roles of Micronutrients

Iron: Iron plays an important role in chlorophyll synthesis, without being part of its molecule. Furthermore, it participates in the molecule of Fe-proteins catalase, cytochrome a, b, c, hyperoxidase *etc.* In addition to that, it is found in the enzymes nitric and nitrate reductase, which are responsible for the transformation of NO_3^- into NH_4^+, in the reduction of SO_4^{2-}, as well as in nitrogenase, which is the responsible enzyme for the atmospheric N capturing [3]. The most characteristic symptom of Fe deficiency is chlorosis of plants. Iron is also constituent of hemo-proteins and nonheme iron proteins, dehydrogenases and ferredoxins. It is involved in photosynthesis, N_2 fixation and valence changes [1].

Manganese: Manganese is part of the molecule of complex of photosystem II (PSII), of the isoform of superoxide desmutases (SODs) MnSOD, and of acid phosphatases. In addition, Mn is activator of the enzymes of carbohydrates metabolism, those of Krebs cycle, as well as of some other enzymes, such as cysteine desulphydrase, glutamyl transferase *etc.* [3, 4].

Of great importance is also the key-role Mn plays in PSII of photosynthesis, and particularly in the reactions liberating O_2, a process, which is crucial in order to be maintained the aerobic life in our planet. In chloroplasts, 3 sources of bound Mn can be distinguished: i) the weakly bound Mn, ii) the strongly bound and iii) the very strongly bound (structural Mn). The weakly bound Mn does not have a functional role in photosynthesis, the strongly bound is the functional one, *i.e.*, that part of chloroplast Mn which is involved in the photosynthetic liberation of oxygen, while the very strongly (structural Mn) one plays an important role in the maintenance of chloroplasts' structure [5]. The exact molecular mechanism of the oxidation of H_2O in PSII still remains a mystery; however, it is known that the complex of oxygen liberation, which is part of PSII and where the oxidation of H_2O occurs, has in its molecule 4 atoms of Mn and one atom of Ca. Recent advances in crystallography revealed the binding of 7 amino-acids in the aggregate of Mn_4Ca; nevertheless, its structural role is not yet fully understood [6, 7] and it still remains a challenge for the scientists.

SODs are a group of metalloenzymes, widely distributed in biological systems, which catalyzes the transformation of reactive oxygen species (ROS) to H_2O_2, according to the reaction: $O_2^- + O_2^- + 2H^+ \rightarrow H_2O_2 + O_2$ (catalyzed by SOD). Previously, ROS are usually produced through the addition of one electron to the molecule of oxygen, according to the reaction $O_2 + e^- \rightarrow O_2^-$. So, in the second stage ROS are transformed into H_2O_2 (the reaction is catalyzed by SOD, as previously mentioned) and finally, H_2O_2 is transformed ('broken') to $2H_2O + O_2$, according to the reaction $2H_2O_2 \rightarrow 2H_2O + O_2$ [8]. The isoforms of SOD can be divided into three categories: i) isoform of SOD with Mn (MnSOD), ii) isoform of SOD with Cu (CuSOD) and iii) isoform of SOD with Fe (FeSOD). The position of each one of these isoforms of SOD inside plants and within the different vegetative organs depends on plant species; for example, in wheat and tomato plants CuSOD and ZnSOD have been detected in the chloroplasts, while MnSOD has been detected in the vacuoles [9, 10]. In tobacco plants MnSOD has been detected in chloroplasts [11].

Another group of enzymes that contains Mn in their molecules are acid phosphatases. Acid phosphatases catalyze the hydrolysis of phosphoric monoesters under acid conditions and have been detected in a great variety of vegetative species, such as potato and rice plants *etc.* The most studied phosphatase is that detected in *Ipomoea batatas* plants, which is a protein with molecular weight 110 KDa, containing two atoms of Mn in its molecule [5].

Manganese acts as activator for about 35 enzymes, catalyzing oxidation and reduction reactions, decarboxylations, hydrolyses *etc.* [3, 5]. Table **1** shows a list of enzymes that are activated by Mn. Manganese is generally involved in the metabolism of carbohydrates and N, in the Krebs cycle reactions, in photosynthesis, in the biosynthetic way of scikimic acid, in the activity of the enzymes oxidase of auxin (IAA oxidase), polyphenol oxidase (PPO), allantoate amylohydrolase *etc.* In the reactions of glycolysis (the process by which glucose is converted to pyruvic acid, together with production of ATP and NADH+H$^+$), Mn is involved in the 1st, 7th, 9th, and 10th (final) stage of the series of reactions [5]. Concerning the influence of Mn on N metabolism, this could be summarized into the following: the most important reaction in N metabolism is the assimilation of NH$_4^+$ to glutamic acid (the product of that reaction is glutamine). That reaction is catalyzed by the enzyme glutamine synthetase, which can use Mn^{2+} as cofactor, instead of Mg^{2+}. Allantoate amylohydrolase is another enzyme that its' activity is influenced by Mn; it is responsible for ureides (product of N metabolism) breakage. This enzyme has an absolute demand on Mn in order to be activated (it can not be activated by another nutrient, such as Mg^{2+}) [8]. In Krebs cycle, a common pathway of the breakage of sugars, proteins and fatty acids, 9 reactions take totally place until the final stage of products. The enzymes participating in this series of Krebs cycle reactions and having absolute demand on Mn in order to be activated are isocitrate and malate dehydrogenase (the enzyme that catalyzes the final reaction of Krebs cycle) [8]. In the photosynthesis of C4 plants, PEP carboxylase (the enzyme that catalyzes the reaction of CO$_2$ assimilation to PEP towards the production of oxaloxic acid) is an enzyme that is activated either by Mn^{2+}, or by Mg^{2+}. However, in the same photosynthetic cycle (that of C4 plants) there is another enzyme, PEP carboxykinase (catalyzing the decarboxylation of oxaloxic acid to PEP and CO$_2$), which has an absolute demand on Mn in order to be activated; in that case, Mg inhibits its activity [8]. Apart from the above mentioned roles of Mn on H_2O photolysis, Krebs reactions, N metabolism, glycolysis *etc.*, it is also involved as cofactor in a series of key-reactions leading to the synthesis of metabolites, highly important for plants. The total of these reactions constitute the metabolic pathway of shikimic acid. Some of the most important products produced as intermediate metabolites in that series of reactions are many kinds of phenols, such as caffeic, chlorogenic, ferrulic acid *etc.* These phenols have been associated with increased resistance of plants to fungi and insect attacks [5]. Lignins and cumarins are also other products associated with increased resistance to microorganism attacks and their production inside plant organs highly depends on Mn mineral nutrition [12, 13]. Huber and Wilhelm (1988) [14] found that young plantlets of *Cucurbita pepo* suffering from Mn deficiency were more prone to attacks by the fungus *Sclerotinia sclerotiorum*, than plantlets receiving adequate Mn nutrition. Finally, Sirkar and Amin (1974) [15] found that Mn deficiency negatively affected auxin (IAA) metabolism because the activity of IAA oxidase significantly increased.

Zinc: Zn plays crucial role in tryptophan biosynthesis, which is the previous stage from IAA (auxin) synthesis (direct influence of Zn on plant growth and biomass production). IAA concentration is significantly reduced in vegetative tissues suffering from Zn deficiency. In addition to the above, Zn is part of some metal-enzymes [3]. Some of these enzymes are anhydrases, dehydrogenases, proteinases and peptidases. Zinc is involved in carbohydrate, nucleic acid and lipid metabolism [1].

Copper: Cu is activator of some enzymes, as well as it is part of enzymes catalyzing oxidation and reducing reactions, such as oxidase of ascorbic acid, lactase, nitrate and nitric reductase *etc.* [3]. Furthermore, Cu is an essential constituent of plastocyanin, a protein which is a component of the electron transport chain of photosystem I in photosynthesis [16], as well as of various oxidases and ceniloplasmin [1]. According to Kabata and Pendias (2001), Cu is involved in oxidation, photosynthesis, protein and carbohydrate metabolism, while it is also possibly involved in symbiotic N_2 fixation and cell wall metabolism [1].

Table 1. Enzymes that are activated by Mn (modified from Burnell, 1988 [5]).

NAD-malic dehydrogenase
NADP-malic enzyme
NAD- isocitric dehydrogonase
NADP-isocitric dehydrogonase
Reductase of hydroxylamine
Hexokinase
Phosphoglyceric kinase
Pyruvic kinase
Pyrophosphorylase
NAD kinase
Adenosine kinase
Arginine kinase
Phosphoglycomutase
Arginase
Alakaline phosphatase
Acid phosphatase
PEP carboxykinase
PEP carboxylase
Enolase, glutamine synthetase, pyruvic carboxylase, Hyperoxidase, MnSOD.

Boron: B plays role in the transfer of sugars along cell membranes, as well as in RNA and DNA synthesis. It also participates to cell division process, as well as to the pectin synthesis [3]. In addition, B is involved in flavonoid synthesis, phosphate utilization and polyphenol production, while it is also constituent of phosphogluconates [1].

Molybdenum: It is part of the enzymes nitrogenase (capturing of atmospheric N), nitric reductase (transformation of NO_3^- to NO_2^-) and molybdoferredoxin. Molybdenum also participates to the metabolism of ascorbic acid, to N_2 fixation, as well as to NO_3^- reduction [1, 3].

Cobalt: It is constituent of cobamide coenzyme, while it is involved in symbiotic N_2 fixation. Cobalt possibly plays also role in nitrogen fixation of non-nodulating plants, and in valence changes stimulation synthesis of chlorophyll and proteins [1].

Diagnostic Tools Used to Assess Micronutrient Deficiency: Symptoms, Critical Concentrations and Fertilization Methods

Although deficiency symptoms of trace elements can not be generalized, some common micronutrient deficiency symptoms (such as depressed plant growth and chlorosis, which is the most frequent symptom) [1], as well as critical micronutrient concentrations for some important crops, are discussed and analyzed. Critical micronutrient concentrations are those below which plant growth is suppressed by 10%. For example, according to Marschner (1995) [8], Mn critical concentrations in mature leaves of plants vary from 10 to 20 mg/kg d.w., depending on vegetal species. According to Dell and Robinson (1993) [17], care must be taken in observing the progress of deficiency symptoms within the shoots over time and in the selection of the correct leaf stage (mature leaves should be selected) for foliar analysis. After fully expanded, mature (physiologically active) leaves have been selected, they are oven-dried in 75°C for 48 hours, they are incinerated and then the ash is dissolved, usually with hydrochloric acid solution. Finally, leaf micronutrient concentrations are determined in atomic absorption spectroscopy and they are compared to critical micronutrient concentrations, established for the same or other plant species. Then, the necessary fertilizer recommendations (soil or foliar ones) should be done, in order to correct micronutrient deficiencies in plants and crops. For example, Somasundaram *et al*. (2011) found that the average Fe concentrations in leaf and petiole tissues of *Citrus* plants were 206 mg/kg dry weight (d.w.) and 161 mg/kg d.w., respectively, which they were considered to be above the normal range (50-100 mg/kg d.w.) [18]. Plant analysis is one of the four methods used to assess nutrient availability or sufficiency and the most expensive one, since it demands laboratory equipment for plant growth; however, it is the most reliable method to determine and interpret the results of leaf micronutrient status. The other three diagnostic tools are: i) the observation of visual symptoms, ii) soil testing and iii) crop growth response. Recently, in order to develop a better diagnostic method, biochemical indicators based on enzymatic assays were proposed [1]. The four approaches mentioned above are becoming widely used, separately or collectively as nutrient availability, or deficiency, or sufficiency diagnostic tools [19].

The observation of visual nutrient symptoms is the cheapest nutritional disorder diagnostic technique, compared to the other three methods. However, it demands a lot of experience on the part of the observer because deficiency symptoms many times are confused with drought, insects and disease infestation, herbicide damage, soil salinity *etc*. In addition to that, sometimes plants maybe on borderline with respect to deficiency and adequacy of a nutrient. In this situation, there are not visual symptoms but the plant is not producing at its maximum capacity. This condition is frequently called 'hidden hunger' [19]. In general, deficiency symptoms caused by immobile nutrients first appear in the upper or younger leaves. The older leaves do not show any symptoms because immobile nutrients do not move from older to younger leaves. This is why we should take mature leaf samples for chemical analyses. The immobile trace elements are Zn, Cu, B, Fe, Mn and Mo. So, the use of visible symptoms in order to assess micronutrient deficiency has the advantage of direct, costless (without laboratory support equipment) field observation, but its basic drawback is that many times it is too late to correct a nutrient deficiency because the disorder is identified when it is too severe to produce visible symptoms [3, 19]. Micronutrient deficiency (Fe, Mn, Zn and Cu deficiency) symptoms in rose crops appeared as growth stunting, light green colour of the leaves, or necrotic patches on leaf blades [17]. According to Somasundaram *et al*. (2011) [18], low and unstable fruit yield, poor quality of fruits and excessive fruit dropping are the main problems provoked possibly by micronutrient deficiencies in a limed *Citrus* sp. crop. The expression of micronutrient deficiency symptoms is not uniform among crops and plant species, as well as among cultivars of the same species. Khoshgoftarmanesh *et al*. (2008) [20] found that the appearance of micronutrient symptoms was different among rose cultivars; more specifically, micronutrient symptoms were absent or only slightly developed in cultivar 'Aqua Fresh'. In contrast, 'Modern Girl' and 'Orange Juice' displayed severe chlorosis and greater reduction in growth. In Table **2** are presented the most typical micronutrient deficiency symptoms, observed in some common plant species.

Table 2. **Symptoms of micronutrient deficiencies in some common plant species (modified by Kabata and Pendias, 2001) [1] and Iannotti [21]).**

Element	Symptoms	Sensitive Crop
B	Chlorosis and browning of young leaves, distorted blossom development, lesions in pith and roots, poor stem and root growth, terminal (end) buds may die	Legumes, *Brassica* sp. (cabbage and relatives), beets and fruit trees (apples and pears)
Cu	Stunted growth, wilting, melanism, disturbance of lignification and of development and fertility of pollen, leaves can become limp, curl or drop.	Cereals (oats), sunflower, spinach, and lucerne (alfalfa)
Fe	Interveinal chlorosis of young organs	*Citrus* sp., grapes, fruit trees (basically peach trees)
Mn	Growth slows, chlorotic or dark spots, young leaves turn pale yellow, often starting between veins, sometimes necrosis of young leaves may occur, leaves, fruits and shoots diminished in size	Cereals, legumes, fruit trees (apples, cherries and *Citrus* sp.).
Mo	Chlorosis of leaf margins, 'fired margin', older leaves yellow, remaining foliage turns light green, leaf deformation (narrow and distorted) due to NO_3 excess	*Brassica* species and legumes
Zn	Interveinal chlorosis, stunted growth, rosette of terminal leaves	Cereals, legumes, grasses, grapes and fruit trees (basically *Citrus* sp.)

Soil test is the most common practice in agricultural soils for making fertilizer recommendations. For food crops soil samples are usually taken from the upper 20 cm of soil profile, since about 80% of their root system remains in this soil volume [22]. After soil samples have been taken, they should be air-dried at root temperature, separated from stones and sieved to pass a 10 mesh screen before analyses. For micronutrient extraction, DTPA is usually used in order to determine micronutrient plant available concentrations. After soil micronutrient concentrations determined, they are compared to critical ones, so fertilizer recommendations can be done in order to restore micronutrient concentrations in sufficient levels. For example, according to Lindsay and Norvell (1978) [23], the critical concentrations of Zn, Fe, Mn and Cu by using the DTPA test are 0.5, 2.5, 1.0 and 0.2 mg/kg dry weight (d.w.) of soil, respectively. However, different critical limits between extractants, as well as between crops for the same extractant, are often found in bibliography. For example, it was found by Katyal (1985) [24] and Takkar *et al.* (1989) [25] that the critical deficiency level for Zn was 0.6 mg/kg soil, respectively, which is in contrast to that referred by Lindsay and Norvell (1978) (0.5 mg/kg d.w.) [23]. Furthermore, the critical Zn levels for corn, according to Soltanpour & Schwab and DTPA methods, were 0.9 and 0.7

mg/kg, respectively [26]. So, it is advised to researchers and agronomists, when they want to compare their results of soil analyses to critical micronutrient deficiency limits established by other researchers, to pay attention to the following points: i) they should compare results obtained under similar soil properties and conditions, ii) they should compare micronutrient extractable concentrations for the same crop (*e.g.*, corn, cotton plants, tomato plants *etc.*), iii) they should compare their results to those of other researchers, determined by using the same extractant (*e.g.*, DTPA), and finally, iv) they should compare their results obtained under field conditions to as many as possible studies conducted by other colleagues and not only to one study, since this may lead to a wrong interpretation of their results.

Visual symptoms, soil and plant analyses are the common practices for identifying nutritional disorders in plants. The best criterion, however, for diagnosing nutritional deficiencies in annual crops is through evaluating crop responses to applied nutrients. If a given crop reacts to an applied nutrient in a soil, this means that this nutrient is deficient for that crop [27]. Relative decrease in yield under nutrient deficiency, as compared to an adequate soil fertility level, can give an idea of the magnitude of nutrient shortage. In addition, biochemical indicators, based on enzymatic assays, were proposed by Rajaratham *et al.* (1974) [28] for assessing B status of the oil palm, Ruszkowska *et al.* (1975) [29] for Cu supply in sunflower leaves, and Gartrell *et al.* (1979) [30] for Cu deficiency in cereals. More specifically, the activity of some 'key' enzymes (enzymes whose activity is influenced by a specific micronutrient deficiency) is mainly correlated with micronutrient levels in plant tissues. However, the practical use of the enzymatic assays is limited due to the high rate of variation and to technical difficulties in the determination of the enzymatic activity [1].

There are two fertilization methods used to correct micronutrient constraints in plants: the first one is soil fertilization and the second is foliar application. Although soil application of fertilizers is more common, convenient and cheaper practice preferred by many farmers, under certain circumstances foliar application may be preferable and more effective. It was found that Zn foliar sprays increased leaf Cu content in maize plants, while the effect of the addition of Zn in soil was not so angle [31]. According to the same author (Aref, 2011), the use of B spraying was more effective on leaf Fe content, relative to the soil B application [31]. Soil application of fertilizers is mainly done on the basis of soil tests, while foliar nutrient applications are mainly done on the basis of visual foliar symptoms or plant tissue tests. Hence, correct diagnosis of nutrient deficiency is fundamental for successful foliar fertilization [19]. Sometimes, foliar fertilization of crops can complement soil fertilization while in other cases it seems to be the unique solution (when soil fertilization does not offer the desired results). One of these cases is when severe Fe deficiency occurs, which is one of the most difficult nutritional constraints to correct, especially on calcareous soils. Foliar spray (0.5-1%) of Fe in the form of $FeSO_4$ solution is the best way to overcome Fe deficiency [18]. According to Habib *et al.* (2012) [32], foliar application with Fe and Zn increased seed yield and its quality in wheat plants, when compared to the control. One of the basic points in foliar fertilizations is the mobility of each necessary trace element within the studied plant species, *i.e.*, micronutrient mobility varies not only between elements, but also among plant species. Papadakis *et al.* (2007) [33] found that Mn and Fe were found to be relatively mobile and strictly immobile nutrients, respectively, within *Citrus* plants (*Citrus aurantium* L. and *C. sinensis* L., cv. Washington navel x *Poncirus trifoliata*) after foliar application with their sulphate salts ($MnSO_4.H_2O$ and $FeSO_47H_2O$). Marschner (1995) [8] reported that Mn and Fe mobility in plant phloem is considered to be low or intermediate, respectively, something which is in contrast to the results of Papadakis *et al.* (2007) [33] for Fe mobility in *Citrus* plants. Generally, micronutrient mobilization rate within plants is complex and large differences exist among various plant species [33].

Foliar uptake is believed to consist of two phases: non-metabolic cuticular penetration, which is generally considered to be the major route of entry, and metabolic mechanisms, which account for nutrient accumulation against a concentration gradient. The second process is responsible for transporting ions across the plasma membrane and into the cell protoplast [1]. The rate by which a chemical substance passes through the cuticle and generally the epidermal tissues of leaves and stems in order to be transported *via* the phloem to the other unsprayed tissues depends on many factors, including anatomical and morphological ones, the concentration and the physicochemical properties of the sprayed substance, the plant species and its nutritional status, as well as the environmental conditions of the agricultural studied area [8, 33]. For example, in the comparison between MnEDTA and $MnSO_4.H_2O$ it was found that the second substance was more efficient than the first one regarding the improvement of the leaf Mn concentration of 'Washington navel' orange trees. In addition to the agronomic advantages, $MnSO_4.H_2O$ was more convenient to be used as spray from a commercial point of view [34]. According to the same authors (Papadakis *et al.*, 2005) the optimum Mn concentration range in $MnSO_4.H_2O$ sprays in order to improve the leaf Mn concentration (within the sufficiency range, *i.e.*, >25mg/kg d.w.) of orange trees varied from 400 to 1200 mg/l (the treatment of 200 mg/l did not present satisfactory results in the improvement of Mn deficiency) [34].

Two kind of soil fertilizers may be used in order to correct micronutrient deficiencies: inorganic chemical fertilizers, or organic amendments (organic chelates of micronutrients with EDTA, EDTPA *etc.*, different kind of manures, branches, leaves through recycling in orchards, composts). Organic amendments may be also used to correct micronutrient deficiencies, although in some cases it was found to increase non-essential element concentrations in soils and plants. For example, according to Demir *et al.* (2010), poultry manure has been traditionally applied to agricultural areas as an organic fertilizer, because it is a good source of organic matter and plant nutrients; in their study, it was found that poultry manure significantly increased fruit yield and Zn (an element required in the human diet) concentrations in tomato plants (*Lycopersicom esculentum* L.). However, leaf Br (a non-essential element for plant growth) concentration was significantly increased at the highest level of applied manure [35]. Application of poultry manure can considerably increase Cu and Zn concentrations in soil because these elements are commonly given as supplements in animal feeds [36]. Generally, elevated concentrations of heavy metals in soils and plants due to the eternal use of manures in agricultural areas may sometimes constitute a great risk for human health. The kind of inorganic chemical fertilizer selected to alleviate a micronutrient deficiency depends on the kind of starvation; *e.g.*, when Mn deficiency occurs it is usually preferred $MnSO_4$, when Zn deficiency is a nutritional problem $ZnSO_4$ may be a good solution; $CuSO_4$ is usually preferred in cases of Cu deficiency *etc.*

According to Westfall *et al.* (2005), by-products of the metal industry have been considerably used as micronutrient fertilizers; $MnSO_4$ (by-product of the hydroquinone manufacture) was an early characteristic example of a by-product of the metal industry used as micronutrient fertilizer in agriculture [37]. However, the danger by the use of these by-products is that, since they may contain heavy metal contaminants (such as As, Cd, Cr, Hg, Pb and Ni), soil application of them may lead to unacceptable levels, which could possibly lead to toxicities in plants and animals, or result in the entry of these metals into the human food chain [37]. In Table **3** they are presented the concentrations of Zn, Cd, Ni and Pb in different Zn fertilizers applied to soils, according to Mordvedt (1985) [38]. Baghouse dusts and flue dusts from Zn smelters and other industries may be a source of Zn fertilizers; because Zn-O by-products are generally dusty and difficult to handle, they are partially acidulated with H_2SO_4 to form granular products called Zn oxy-sulfates, containing 20-50% Zn [38]. Westfall *et al.* (2005) [37] refer that Zn availability in granular fertilizers was related to water solubility ($r^2=0.92$) and not to total Zn content. Furthermore, according to the same authors, the relative availability coefficients for organic and inorganic Zn fertilizers was also highly

Table 3.   Concentrations of Zn, Cd, Ni and Pb in Zn fertilizers applied to soil [38].

Zn Fertilizer	Total Zn (%)	Metal Concentration (mg/kg)		
		Cd	Ni	Pb
$ZnSO_4$ (reagent)	27.5	1	1	1
$ZnSO_4$-1	34.0	2,165	92	60
Zn oxysulfate-1	35.1	590	158	44,000
Zn oxysulfate-2	41.5	1,970	19	400
ZnO (reagent)	73.4	1	1	1
ZnO by-product-1	58.3	243	8,950	1,900
ZnO by-product-2	34.0	1,420	250	52,000

related to water solubility and independent of total Zn content. Amrani *et al.* (1999) [39] found that a Zn fertilizer should contain at least 50% water soluble Zn to adequately supply Zn for plants. Fe and Zn chelates are also used in cases of Fe and Zn deficiencies, respectively. According to Somasundaram *et al.* (2011) [18], the most reliable means of correcting Fe chlorosis is by soil application of Fe chelates. Information in detail on the fertilizers used in each micronutrient deficiency is given in the relevant chapters describing these deficiencies.

Finally, some micronutrient input-output budgets in different ecosystems (agricultural or forest) are presented below. Unfortunately, the input-output balance for trace elements is rarely met, even under natural conditions [1]. The basic input sources for agricultural ecosystems are fertilizers, different kind of manures, fungicides, insecticides, the parent material from which soils are originated from, the irrigation water (for example, some waters used for irrigation in Greece are rich in B or Mn), atmospheric deposition (trace metals loading of the agro-ecosystem, especially in agricultural areas which are close to mines, industrial zones *etc.*, or acid rains in industrial polluted areas which lead to soil acidity, thus increasing in an indirect way trace element solubility and uptake by plants), while the output ones are yields of orchards and crops, pruning and leaching due to unfavourable soil conditions (high acidity, especially when it is combined with light texture *i.e.*, sandy soils, and/or high redox potential (E_h) values of up to -300 mV under anaerobic conditions). According to Kabata and Pendias (2001), heavy soils provide good storage for trace elements and will supply them at a slow rate; light soils on the other hand can be a source of easily available trace elements during a short period of time, but they have the drawback to loose their pool of available micronutrients at a quite high rate due to leaching [1]. The same authors also refer that leaching of trace elements was found to be higher than atmospheric input only in acid forest soils. In Table **4** are presented the amounts of micronutrients removed by good yields of some crop species. As it is clear from these data, the amount of the same trace element removed by yields significantly differed between plant species, so this is why different fertilization practices (application and doses) should be adopted in order to cover the loses from yields. Finally, Table **5** shows the budgets of Cd in some agricultural soils of Poland and Germany. It is clear that, in both countries, the Cd balance (input-output) was positive; furthermore, the basic input Cd sources were fertilizers, sludges, harvest residues through the process of recycling, as well as some atmospheric depositions, while the basic output ones were yields removal and seepage water.

Table 4.   Amounts of some micronutrients removed by good yields of various crops [40].

Crops Harvested and Portion Used for Analysis		Micronutients Removed (kg/ha)						
		Yield Level t/ha	Chlorine (Cl)	Boron (B)	Copper (Cu)	Iron (Fe)	Manganese (Mn)	Zinc (Zn)
Alfalfa	- hay	1.3	6	0.10	< 0.1	0.20	0.70	0.70
Barley	- grain	4.0	8	0.10	< 0.1	0.30	0.10	0.10
	- straw	-	1	0.02	< 0.1	0.01	0.70	0.10
Corn	- grain	9.5	2	0.70	< 0.1	0.20	0.10	0.20
	- stover	-	1	0.06	< 0.1	1.0	1.70	0.30
Oats	- grain	4.0	1	-	< 0.1	1.00	0.20	0.10
	- straw	-	1	-	< 0.1	0.20	0.20	0.40
Peas	- vines & pods	-	-	0.07	< 0.1	0.70	0.50	0.10
Potatoes	- white, tubers	40	27	0.07	< 0.1	0.90	0.20	0.10
Wheat	- grain	4.0	6	0.06	< 0.1	0.50	0.20	0.20
	- straw	-	2	0.02	< 0.1	0.20	0.30	0.10

Table 5.   Budgets of Cd in agricultural soils of Poland and Germany (g/ha/year) [1].

Input/Output Sources	Poland	Germany
Input		
Fertilizers	1-2.5	1-6
Slurry	2.5	-
Sludges	1.5	<1-25
Harvest residues (recycling)	3	0.3-8
Atmospheric input	2.5-4	3-8
Output		
With crops	3	1-5
With seepage water	3	1-2
Total output	6	2-7
Balance (net gain)	4.5-7.5	1.3-40

CONCLUSION

Despite the fact that micronutrients are found in traces (very low concentrations, p.p.m.) in soils and plants, they are as essential for plant metabolism and growth as macronutrients, *i.e.*, they are part of enzymes or constituent of other molecules, or they activate other enzymes, which they also regulate very important physiological functions. Many trace elements are part of enzymes or constituent of other molecules. More specifically, Fe participates in Fe-proteins catalase, cytochrome a, b, c, hyperoxidase, Mn is part of the molecule of complex of photosystem II, of the isoform of superoxide desmutases MnSOD, and of acid phosphatases. Zinc participates in different anhydrases, dehydrogenases, proteinases and peptidases, Cu is an essential constituent of plastocyanin, a protein which is a component of the electron transport chain of photosystem I, Mo is part of the enzymes nitrogenase (capturing of atmospheric N), nitric reductase (transformation of NO_3^- to NO_2^-) and molybdoferredoxin, Co is constituent of cobamide coenzyme *etc.* Trace elements are also activators of many enzymes (especially Mn which was found to be activator for more than 35 enzymes), or they are involved in an indirect way in different physiological functions, such as photosynthesis, carbohydrates, lipids and N metabolism, atmospheric N fixation, cell division process *etc.*

There are 4 different ways to detect and assess micronutrient constraints in plants: i) the observation of visual symptoms, ii) plant tissue analysis, iii) soil testing, and iv) crop growth response. Recently, more and more are used the biochemical indicators as early micronutrient deficiency detectors. There are advantages and disadvantages between these methods: visual observation of symptoms is a costless method (it does not demand any laboratory equipment, as in the other methods), but many times, after symptoms appeared, plant metabolism has already been disrupted, so it may be too late to solve the problem. Furthermore, it demands a lot of experience. Plant tissue analysis and soil testing are more reliable than visual observation and generally they are the most common practices used to detect micronutrient deficiencies in agriculture. In annual crops, plant growth responses to applied nutrients give very good results. There are two fertilization methods used to correct micronutrient constraints in plants: the first one is soil fertilization and the second is foliar application. Although soil application of fertilizers is more common and cheaper practice preferred by many farmers, under certain circumstances foliar application may be more economic and more effective. In soil, two kinds of fertilizers may be used in order to correct micronutrient deficiencies: inorganic chemical fertilizers, or organic amendments (different kind of manures, branches, leaves through recycling in orchards, composts *etc.*).

REFERENCES

[1]   Kabata-Pendias A, Pendias H. Trace Elements in Soils and Plants. 3rd ed. CRC Press: USA 2001.

[2]   Frerot H, Petit C, Lefebvre C, Gruber W, Collin C, Escarre J. Zinc and Cadmium accumulation in controlled crosses between metallicolous and non-metallicolous populations of *Thlaspi caerulescens* (Brassicaceae). New Phytol 2003; 157: 643-648.

[3]   Therios I. Mineral Nutrition and Fertilizers. Dedousis Publications, Thessaloniki, Greece 1996; pp. 174-177. (In Greek).

[4]   Voyiatzis D, Petridou M. Biology of Horticultural Plants. Gartaganis Publications, Thessaloniki, Greece 2003; pp. 89-93. (In Greek).

[5]   Burnell JM. The biochemistry of manganese in plants. In: Graham RD, Hannam RJ, Uren NC, Eds. 'Manganese in soils and plants', Proceedings of the International symposium on 'Manganese in soils and plants'. Kluwer Academic Publishers, 1988; pp. 125-137.

[6]   Kusunoki M. 2007. Mono-manganese mechanism of the photosystem II water splitting reaction by a unique Mn_4Ca cluster. Biochimica et Biophysica Acta 2007; 1767: 484-492.

[7]   Nagata T, Nagasawa T, Zharmukhamedov SK, Klimov VV, Allakhverdiev SI. Reconstitution of the water-oxidizing complex in manganese-depleted photosystem II preparations using synthetic binuclear Mn(II) and Mn (IV) complexes: production of hydrogen peroxide. Photos Res 2007; 93: 133-138.

[8]   Marschner H. Mineral Nutrition of Higher Plants, 2nd edition. Academic Press, London 1995; pp. 324-333.

[9]   Wu G, Wilen RW, Robertson AJ, Gusta LV. Isolation, chromosomal localization, and differential expression of mitochondrial manganese superoxide dismutase and chloroplastic copper/zinc superoxide dismutase genes in wheat. Plant Physiol 1999; 120: 513-520.

[10]   Shenker M, Plessner OE, Tel-Or E. Manganese nutrition effects on tomato growth, chlorophyll concentration, and superoxide dismutase activity. J Plant Physiol 2004; 161: 197-202.

[11]   Yu Q, Osborne LD, Rengel Z. Increased tolerance to Mn deficiency in transgenic tobacco overproducing superoxide dismutase. Ann Bot 1999; 84: 543-547.

[12]   Rengel Z, Pedler JF, Graham RD. Control of Mn status in plants and rhizosphere: genetic aspects of host and pathogen effects in the wheat take-all interaction. In: Manthley JA, Crowley DE, Luster DG, Eds. Biochemistry of metal micronutrients in the rhizosphere. Lewis Publishers/CRC Press, Boca Raton, FL, USA 1994; pp. 125-145.

[13]   Santandrea G, Schiff S, Bennici A. Effects of manganese on *Nicotiana* species cultivated *in vitro* and characterization of regenerated Mn-tolerant tobacco plants. Plant Sci 1998; 132: 71-82.

[14]   Huber DM, Wilhelm NS. The role of manganese in resistance to plant diseases. In: Graham RD, Hannam RJ, Uren NC, Eds. 'Manganese in soils and plants', Proceedings of the

International symposium on 'Manganese in soils and plants'. Kluwer Academic Publishers, 1988; pp. 155-173.

[15]   Sirkar S, Amin JV. The manganese toxicity of cotton. Plant Physiol 1974; 54: 539-543.

[16]   Losak T, Hlusek J, Martinec J, *et al.* Nitrogen fertilization does not affect micronutrient uptake in grain maize (*Zea mays* L.). Acta Agric Scand Section B-Soil Plant Sci 2011; 61: 543-550.

[17]   Dell B, Robinson M. Symptoms of mineral nutrient deficiencies and the nutrient concentration ranges in seedlings of *Eucalyptus maculata* Hook. Plant Soil 1993; 155/156: 255-261.

[18]   Somasundaram J, Meena HR, Singh RK, Prasad SN, Parandiyal AK. Diagnosis of micronutrient imbalance in lime crop, in semi-arid region of Rajasthan, India. Com Soil Sci Plant Anal 2011; 42: 858-869.

[19]   Fageria NK, Barbosa-Filho MP, Moreira A, Guimaraes CM. Foliar fertilization of crop plants. J Plant Nutr 2009; 32: 1044-1064.

[20]   Khoshgoftarmanesh AH, Khademi H, Hosseini F, Aghajani R. Influence of additional micronutrient supply on growth, nutritional status and flower quality of three rose cultivars in a soilless culture. J Plant Nutr 2008; 31: 1543-1554.

[21]   Iannotti M. Plant Nutrient deficiencies: Identifying Plant problems. http://gardening.about.com/od/gardenproblems/a/NutrientDeficie.htm

[22]   Fageria NK, Baligar VC, Clark RB. Physiology of Crop Production. New York: The Haworth Press 2006.

[23]   Lindsay WL, Norvell WA. Development of a DTPA soil test for zinc, iron, manganese and copper. Soil Sci Soc Am J 1978; 42: 6421-6428.

[24]   Katyal JC. Research achievements of all India coordinated scheme of micronutrients in soils and plants. Fertilization News 1985; 30: 67-80.

[25]   Takkar PN, Chhibba IM, Mehta SK. Twenty years of coordinated research on micronutrients in soils and plants. Bull Ind Inst Soil Sci 1989; 1: 76.

[26]   Martens DC, Lindsay WL. Testing Soils for Copper, Iron, Manganese and Zinc. In: Westerman RL, Ed. Soil Testing and Plant Analysis, 3rd Edition. Soil Sci Soc Am. Madison, WI: USA 1990; p. 229.

[27]   Fageria NK, Baligar VC. Nutrient availability. In: Hillel D, ED. Encyclopedia of Soils in the Environment. San Diego, CA, Elsevier, 2005; pp. 63-71.

[28]   Rajaratham JA, Lowry JB, Hock LI. New method for assessing boron status of the oil palm. Plant Soil 1974; 40: 417.

[29]   Ruszkowska M, Lyszcz S, Sykut S. The activity of catechol oxidase in sunflower leaves as indicator of copper supply in plants. Pol J Soil Sci 1975; 8: 67.

[30]   Gartrell JW, Robson AD, Loneragan JF. A new tissue test for accurate diagnosis of copper deficiency in cereals. J Agric West Aust 1979; 20: 86.

[31]   Aref F. Influence of zinc and boron nutrition on copper, manganese and iron concentration in the maize leaf. Aus J Basic Appl Sci 2011; 5: 52-62.

[32]   Habib M. Effect of supplementary nutrition with Fe, Zn chelates and urea on wheat quality and quantity. Afr J Biotech 2012; 11: 2661-2665.

[33]   Papadakis IE, Sotiropoulos TE, Therios IN. Mobility of iron and manganese within two *Citrus* genotypes after foliar applications of iron sulphate and manganese sulphate. J Plant Nutr 2007; 30: 1385-1396.

[34]   Papadakis IE, Protopapadakis E, Therios IN, Tsirakoglou V. Foliar treatment of Mn deficient 'Washington navel' orange trees with two Mn sources. Sci Hortic 2005; 106: 70-75.

[35]   Demir K, Sahin O, Kadioglu YK, Pilbeam DJ, Gunes A. Essential and non-essential element composition of tomato plants fertilized with poultry manure. Sci Hortic 2010; 127: 16-22.

[36]   Uprety D, Hejcman M, Szakova J, Kunzova E, Tlustos P. Concentration of trace elements in arable soil after long-term application of organic fertilizers. Nutr Cycl Agroecosyst 2009; 85: 241-252.

[37]   Westfall DG, Mordvedt JJ, Peterson GA, Gangloff WJ. Efficient and environmentally safe use of micronutrients in agriculture. Com Soil Sci Plant Anal 2005; 36: 169-182.

[38]   Mordvedt JJ. Plant uptake of heavy metals in zinc fertilizers made from industrial by-products. J Environ Quality 1985; 14: 424-427.

[39]   Amrani M, Westfall DG, Peterson PA. Influence of water solubility of granular zinc fertilizers on plant uptake and growth. J Plant Nutr 1999; 22: 1815-1827.

[40]   McKenzie RH. Micronutrient requirements of crops. Alberta: Agriculture and Rural Environment; 1992: http://www1.agric.gov.ab.ca/$department/deptdocs.nsf/all/agdex713

CHAPTER 3

Iron Deficiency

Abstract: Despite the fact that globally Fe is in great abundance and total Fe content in soils is high, many times plants suffer from Fe chlorosis. This happens because the greatest part of this content exists in insoluble forms (oxides and hydroxides of Fe, phosphate substances of Fe *etc.*), thus it can not be taken up by plants. Iron solubility and uptake depends on many soil (pH, soil humidity, C.E.C., organic matter, $CaCO_3$ content *etc.*) and non-soil factors (such as root exudates, plant-microbial interactions, production of phytosiderophores, root ferric reductase activity, fertilization, grafting on Fe-tolerant rootstocks, crop management practices *etc.*). There are two mechanisms adopted by plants in order to take up Fe from soil: strategy I, and strategy II. Strategy I is a complex Fe uptake mechanism developed by all plants, with the exception of *Poaceae* plants, which belong to strategy II. Strategy I uptake mechanism is based on the reduction of external Fe^{3+} to Fe^{2+} through the induction of Fe^{3+} chelate reductase enzyme. Strategy II uptake mechanism is based on an increase in the synthesis and secretion of phytosiderophores (PS) to the environment of the rhizosphere. Then, the PS-Fe complexes are easily taken up by plants.

Many horticultural and agronomic crops (such as apple, grape, peach and *Citrus*), which belong to strategy I species, are sensitive to Fe deficiency. From strategy II species, rice and sorghum are among the most sensitive crops. Since Fe is involved in chlorophyll synthesis, chlorophyll content and photosynthetic rate, they are usually decreased under Fe deficiency; this is the reason why chlorosis is the most usual macroscopic symptom observed under conditions of Fe starvation. There are many mechanisms of tolerance adopted by plants in order to face Fe deficiency, like enhanced ability to induce H^+ extrusion in strategy I plants, production of greater quantities of PS that are released by roots in order to mobilize Fe in strategy II plants, modification of the morphology of their root system in order to increase Fe uptake *etc.* Finally, there are two basic methods of supplying Fe in plants: through soil and foliar application; the foliar application is very advantageous under alkaline soil conditions. There are two basic categories of Fe fertilizers: the inorganic ones, based on inorganic Fe compounds, such as Fe salts (*e.g.*, $Fe(SO_4)7H_2O$) and insoluble compounds, such as Fe oxides-hydroxides and the organic fertilizers, based on organic compounds, like Fe-EDTA and Fe-EDDHA.

All the above mentioned topics concerning soil and plant factors influencing Fe solubility and uptake, strategies of Fe uptake, mechanisms of tolerance adopted by plants in order to face Fe starvation, as well as methods of fertilizer application, and substances used to alleviate chlorosis and organic fertilizers, are fully analyzed in this chapter under the light of the most recent and important scientific papers.

Keywords: Fe chelates, Fe chlorosis, Fe deficiency, Fe-DTPA, Fe-EDTA, Fe solubility, Fe-HEEDTA, $FeSO_4$, Fe tolerance, strategy I, strategy II.

Theocharis Chatzistathis

INTRODUCTION

Iron is one of the major constituents of the lithosphere and comprises approximately 5%, being concentrated mainly in the mafic series of magmatic rocks. However, the global abundance of Fe is calculated to be around 45% [1]. Iron availability for plants depends greatly on weathering and solubility. The basic sources of soluble Fe^{2+} in soils are the different Fe-containing minerals from which Fe^{2+} is 'liberated' through the process of weathering. The most important Fe-containing minerals are hematite, magnetite, goethite *etc.* Divalent Fe (Fe^{2+}) may then be converted to trivalent form (Fe^{3+}, non available for plants); generally, well-aerated, oxidizing and alkaline soil conditions promote the precipitation of Fe, whereas acid and reducing conditions promote Fe solubility (the predominance of Fe^{2+}). Other factors that influence Fe solubility in soils and uptake by plants are pH, the presence of some bacteria genus, such as *Thiobacillus* and *Ferrobacillus*, root exudation and reducing capacity of Fe^{3+} to Fe^{2+}, the content in phosphoric substances, concentrations of other micronutrients and metals, such as Mn, Zn, Cu and Ni, Cd *etc.* [2, 3].

Iron deficiency is the most prevalent micronutrient constraint of many crops when grown on calcareous and alkaline soils [4]. The most characteristic symptom is chlorosis of young leaves, since it is a relatively immobile nutrient inside plants, while poor yields and reduced nutritional quality of food are also the other two most found abnormalities of Fe deprivation. Of course, a great variability in the intensity of Fe deficiency symptoms, in the range of critical deficient concentrations, as well as in tolerance (as that assessed by chlorosis and growth parameters), are usually observed not only within plants species, but also among cultivars of the same species.

Increasing Fe available levels in major staple food crops is an important strategy to reduce Fe deficiency in people [5]. About two million people in the world are anaemic, mainly due to Fe deficiency [6]. For that purpose, in calcareous and alkaline soils, in order to overcome Fe deficiency in plants it is preferred either the use of controlled release fertilizers [7], or the foliar fertilization. Some of the most important substances used as foliar fertilizers to face Fe deficiency are $FeSO_4$, Fe-DTPA, Fe-HEEDTA *etc.* [2].

The purposes of this chapter are: i) to fully present and analyze all the factors influencing Fe solubility in soils and Fe uptake by plants, as well as the sources of Fe in soils (the basic Fe-containing minerals and rocks), ii) to give a thorough description of symptoms, as well as to present some indicative critical Fe deficient concentrations in different plant species (fruit trees, forest species, annual species *etc.*), iii) to provide the reader with some very useful information concerning the selection of tolerant genotypes to Fe deficiency [8]. In addition, within the purposes of the present chapter are also: iv) to describe some tolerance mechanisms adopted by plants in order to face Fe deficiency (adaptive strategies), such as increased acidification of the rhizosphere (which causes reduction of Fe^{3+} to Fe^{2+}), and greater production and accumulation in the root system of organic acids and phenolics [9]. Finally, v) the properties, the advantages and disadvantages of Fe fartilizers, as well as the indicated quantities per plant used by some farmers and researchers in different plant species is also another very important section of this chapter. Apart from the fertilization practices that should be adopted to overcome Fe deficiency, vi) the exploitation of the plants' potential for Fe mobilization and utilization through breeding, transgenic approaches [10] and crop management systems [11] are also very important practices discussed in this chapter.

Fe-Containing Minerals and Rocks and Fe Content in Soils

With 5% of the earth's crust, Fe is the fourth most abundant element in the geosphere, only inferior to oxygen, silicon and aluminium. The Fe content of soils varies from 1 to 20%, averaging 3.2%, but its normal concentration in plants is only 0.005% [12]. The main reason is that Fe in soils exists mainly in the forms of hydrogen oxide, oxide, phosphate and other deposited compounds [13], which are insoluble Fe soil forms. It is known that the chemical composition of parent material has a direct effect on chemical properties of soil [2]. In rich in Fe parent materials, it is expected a high soil Fe content. However, elevated total Fe content of soils does not necessarily guarantee a high uptake by plants, since Fe solubility is influenced by various soil and other factors, which will be analyzed in detail in the next paragraph.

Among the most important Fe-containing minerals are those of hematite, magnetite, biotite, siderite, goethite, granodiorite, pyrite, vivianite, chromite, cubanite, germanite *etc.* [14, 15]. According to Arrieta and Grez (1971), who studied the solubilizing action of 28 strains of microorganisms on different Fe-containing minerals isolated from soils of Chile, the solubility of Fe in these minerals depended on their nature, crystalline structure, the concentration of metabolic products, or all these three factors together [14].

The critical Fe level in soils by using DTPA solution is 4.5 mg/kg d.w. [16]. Irmak *et al.* (2008) [17] found that Fe extractable concentrations in calcareous soils of Cukurova region in Turkey varied from 2.60 to 6.0 mg/kg d.w. in 2005 and from 6.96 to 12.70 mg/kg d.w. in 2006, so an important part of soil samples received was found to be below the critical Fe limit [17].

Factors Influencing Fe Solubility in Soils and Uptake by Plants

Iron in soils exists as Fe^{3+} and Fe^{2+}, although it is basically taken up as Fe^{2+} by plants. So, each factor (soil, plant or other one) which increases Fe solubility enhances Fe uptake. The factors influencing Fe solubility in soils can be divided into three categories: a) soil factors, b) plant factors and c) factors influenced by human activities. In the first category (soil factors) are included factors such as pH, $CaCO_3$ content, organic matter, cation exchange capacity (C.E.C.), soil moisture, oxygen and temperature, salt concentration, the content of phosphoric ions, the concentration of other micronutrients and heavy metals, the presence of some bacteria genus, such as *Thiobacillus* and *Ferrobacillus*, the content of Fe oxides and hydroxides *etc.* In the second category (plant factors) are included factors such as rhizosphere acidification (release of H^+), root exudation (root release of phytosiderophores, organic acids *etc.*), mycorrhiza *etc.* Strategies I and II adopted by plants in order to enhance Fe uptake are also described in this subsection. Finally, in the third category (factors influenced by human activities), factors such as soil erosion, intensification of agricultural practices, crop management practices adopted by the farmers, as well as the kind and the quantities of N and P fertilizers used by farmers should not be omitted from this analysis. All these factors belonging in one of the three categories mentioned above are fully analyzed below.

Soil Factors

I. **pH.** It is one of the most important factors influencing Fe availability in soils and uptake. Iron is the micronutrient, which is mostly influenced by pH [2]. Generally, Fe solubility is decreased with the decrease of soil acidity (*i.e.*, with the increase of soil pH); for each unit pH increases, Fe^{2+} solubility is decreased 100 times [18]. Iron chlorosis is a major nutritional disorder in crops growing in calcareous soils. According to Fodor *et al.* (2012) [19], as pH increased from 4.5 to 7.5, the root ferric chelate reductase (FCR) activity of cucumber plants significantly decreased. Soluble inorganic Fe salt applications to soil are usually quite inefficient in calcareous soils due to the rapid transformation of most of the Fe applied into highly insoluble compounds, such as Fe (III) hydroxides. This occurs even when very high doses of these low cost Fe-fertilizers are applied [20].

II. **High CaCO₃ Content.** Iron deficiency is the most common micronutrient constraint provoked by high bicarbonate content. Under such soil conditions Fe chlorosis occurs. Under calcareous conditions the total Fe content in soils may be high, but the available fraction for plants is insufficient [21]. This happened due to the low solubility of Fe oxides at the alkaline pH conditions that are buffered by the presence of high bicarbonate content in these soils [22]. Apart from the indirect effect of high bicarbonate content (due to the increased pH) on Fe availability, the direct effect (due to the antagonism between increased Ca^{2+} concentration and Fe^{2+}) should not be omitted as responsible for the decreased Fe uptake under calcareous soil conditions. Antagonism between Ca^{2+} and Fe^{2+} ions is also the conclusion of the study of other researchers; according to Kabata and Pendias (2001), the antagonistic effect of Ca on Fe uptake is very complex and is related to both growth medium and intracellular metabolism [1].

III. **Cation Exchange Capacity (C.E.C.).** The ability of the solid phase to exchange cations, the so called C.E.C., is one of the most important soil properties governing the cycling of trace elements in soil [1]; Fe is one of these trace elements, as it is positively charged (Fe^{2+}). Since the colloid of clay surfaces are negatively charged, they have the ability to exchange Fe^{2+}

cations with other ones chemically equivalent from the soil solution, so Fe^{2+} may be available for uptake in soil solution. The richer is one soil in clay and organic matter content, the higher C.E.C. values it has.

IV. **Organic Matter.** Iron mobility and solubility are increased after the formation of organic complexes. Generally, soil Fe exhibits a great affinity to form mobile organic complexes and chelates. These compounds are largely responsible for Fe migration between soil horizons and for Fe leaching from soil profiles and they are also important in the supply of Fe to plant roots [1].

V. **Soil Moisture (Flooded-Anaerobic or Aerobic Conditions).** Under flooded (anaerobic) conditions Fe^{3+} is converted to Fe^{2+} due to the lack of soil oxygen. In these cases, Fe availability for plants is significantly increased; according to Fan *et al*. (2012), with increasing water shortages in China rice cultivation is gradually shifting away from continuously flooded conditions to partly or even completely aerobic ones; this will increase, according to the same authors, Fe deficiency in rice and will deteriorate the problem of Fe starvation in humans, who depend on rice for their nutrition [23]. Iron deficiency is often a nutritional problem for crops in upland soils and it is one of the primary limitations for rice production in aerobic areas [24].

VI. **Interaction with Other Nutrients.** From the interaction effects between Fe and other nutrients, the most important is probably that between Fe and Mn. Indeed, many researchers have found that under Mn toxicity Fe uptake by plants is suppressed [25, 26]. This is ascribed to the antagonism between Mn^{2+} and Fe^{2+}. Nevertheless, in some cases the influence of Mn^{2+} on Fe^{2+} uptake may be synergistic. Chatzistathis (2008) found that under excess Mn conditions Fe uptake was significantly suppressed in olive cultivars 'Picual', 'Koroneiki' and 'FS-17', while it was enhanced in cultivar 'Manaki' [26]. The adverse influence, *i.e.*, that of Fe on Mn uptake, is also of great importance. According to Ghasemi-Fasaei *et al*. (2005), the suppression effect of Fe on plant Mn concentration in *Cicer arietinum* L. (var. Pars) plants was not due to the reduction in root: shoot ratio, dilution effect or reduction of Mn uptake by root, but due to the antagonistic effect of Fe on the translocation of Mn from root to shoot [4]. Finally, Fe uptake is many times significantly suppressed due to rich P fertilizations. In that case, insoluble phosphate Fe substances are usually formed [2].

VII. **The Presence of Fe Oxidizing or Reducing Bacteria.** It was found that five moderately thermophilic iron-oxidizing bacteria, including representative strains of the three classified species (*Sulfobacillus thermosulfidooxidans*, *Sulfobacillus acidophilus*, and *Acidimicrobium ferrooxidans*), were shown to be capable of reducing ferric (Fe^{3+}) to ferrous (Fe^{2+}) iron when they were grown under oxygen limitation conditions [27].

Plant Factors

I. **Root Exudates.** In some strategy I plant species the release of organic compounds, such as phenolics, flavins, sugars and organic acids, could help in the solubilization of Fe-containing compounds [28]. In two barley (strategy II plant species) cultivars (*Hordeum vulgare* L., cvs. Steptoe and Morex), root extracts of plants, grown under Fe-deficient conditions, showed higher activities of enzymes related to organic acid metabolism, including citrate synthase, malate dehydrogenase and phosphoenolpyruvate carboxylase, compared to the activities measured in root extracts of Fe-sufficient plants. Furthermore, the concentration of total carboxylates was higher in Fe-deficient roots of both cultivars, with citrate concentration showing the greatest increase [29]. According to Shi *et al.* (2012), root exudates (phytosiderophores) from Fe deficient wheat plants could improve their Fe nutrition in the presence of insoluble $Fe(OH)_3$. Supplying phytosiderophores released by wheat plants significantly increased Fe concentration and content in shoots of aerobic rice plants under insoluble $Fe(OH)_3$ treatments [30]. In roots of *Hordeum vulgare* L., five phytosiderophores (deoxymugineic acid, mugineic acid, epihydroxymugineic acid, avenic acid and hydroxyavenic acid) were identified under Fe deficiency conditions [31].

II. **Morphological, Biochemical and Physiological Adaptations Due to Fe Deficiency.** Under low Fe concentrations, the increase of root hair length allows effective uptake of Fe; this is a strategy adopted by many plant species in order to survive under conditions of low Fe availability. According to Schmidt *et al.* (2003), the increase of root hair length and number of transfer cells are positively correlated with the amount of detectable H^+-ATPases. These morphological and physiological responses in roots under Fe starvation, might be regulated by ethylene and/or auxin signaling [32]. It was also found that ethylene (and possibly salicylic acid) is involved in the regulation of Fe reduction in explants of peach trees [33]. Hence, ethylene may play a role in inducing the morphological characteristics of Fe-deficient plants. Another possible explanation of the role of ethylene in the regulation of Fe reduction in peach trees under Fe deficiency is that it may directly affect transcription or translation of the *FRO2* gene, responsible for the synthesis of the Fe(III) chelate reductase [33]. According to Romera *et al.* (2011), auxin, ethylene and NO increase under Fe deficiency, which would be necessary for the up-regulation of Fe acquisition genes in strategy I plants. Each one influences the production of the other two and all of them require low Fe (probably low phloem Fe) to be effective. Some results suggest that auxin acts upstream of ethylene and NO and that, perhaps, ethylene is the last activator of the Fe acquisition genes [34]. NO was found to improve plant growth, alleviate leaf interveinal chlorosis and increase the activity of Fe(III) reductase activity in peanut plants, when growing on calcareous soils [35]. Mlodzinska (2012) refer that two genes (CsHA2 and CsHA3) were isolated from different parts of cucumber; these two genes were up-regulated under Fe deficiency in cucumber roots, while their expression was decreased under high Fe availability. Nevertheless, the post-translational modification of protein proton-extruding H^+-ATPase and its accumulation as a target of activation in Fe-deficient plants might not be responsible for the increase in the H^+-ATPase activity [36]. According to Vigani *et al.* (2012), a set of genes over-expressed under Fe deficiency, such as those coding for calmodulin; calmodulin was found to be accumulated in Fe deficient root apexes of *Cucumis sativus* L. (typical strategy I plant species) plants [37]. In strategy I plants root tips swell in response to Fe deficiency [38]. However, in the study of Shi *et al.* (2012) it was found that root-tip widths of aerobic rice were similar in –Fe and +Fe treatments, while Fe-deficient cucumber had bigger root tip width, than the Fe sufficient one [30].

III.   **Plant-Microbial Interactions.** The efficiency by which plant roots acquire Fe depends not only on root architecture, protons released and roots exudates, but also on the presence of plant-microbial interactions [39]. Plant-mycorrhiza is the most characteristic example of mutual association, beneficial for both parts (the benefit for the fungus is the supply of organic substances produced by photosynthesis, while the benefit for plant is the increase of nutrient uptake from soil-Fe in our case).

Factors Influenced by Human (Anthropogenic) Activities

I.   **Fertilization.** In the cases of Fe deficiency, Fe fertilizers are used in order to increase Fe availability in soil. However, under calcareous/alkaline conditions, soil application of inorganic source of Fe is not as efficient as leaf fertilization in the correction of Fe deficiency [4]. According to Fuentes *et al.* (2012), natural hetero-ligand Fe (III) chelates (Fe-NHL) (these chelates involve the participation in the reaction system of a partially humified lignin-based natural polymer and citric acid) are able to provide Fe to chlorotic *Citrus* trees, with results comparable to Fe-EDDHA [40]. Apart from the direct effect of Fe fertilization on the increase of Fe availability, some organic Fe fertilizers may favour rooting of plants (thus they may enhance Fe uptake in an indirect way through the increase of root formation); according to Molassiotis *et al.* (2003), who studied the effect of organic (Fe-EDTA and Fe-EDDHA) and inorganic ($FeCl_3$) iron substances on rooting of the rootstock GF-677 (*Prunus amygdalus* x *Prunus persica*) *in vitro*, full rooting (100%) was observed in explants nourished with Fe-EDDHA, while less rooting was found in the presence of $FeCl_3$ and no root formation was observed in explants nourished with Fe-EDTA [41]. Fertilization with Fe not only increases Fe availability and uptake, but also influences its translocation and distribution among plant tissues; indeed, great differences in the uptake and translocation of Fe was found between chlorotic and non-chlorotic sorghum plants receiving different Fe foliar treatments (ferrous sulphate alone, ferrous sulphate + citric acid, ferrous sulphate + thiourea, ferrous sulphate + thioglycollic acid) [42].

II. **Grafting on Tolerant to Fe Chlorosis Rootstocks.** Grafting is a technique that is used by farmers in order to face many agronomical problems. For example, the peach rootstock GF-677 (*Prunus amygdalus* x *Prunus persica*) is a very tolerant to Fe chlorosis rootstock. Furthermore, quince (*Cydonia oblonga* Mill, cv. BA29) may be used as pear (*Pyrus communis* L.) rootstock, since it was found that grafting pear seedlings on semi- woody rooted cuttings of quince resulted in a higher tolerance to Fe deficiency, than when quince plants were not grafted [43].

III. **Increase of phytosiderophore production through gene transfer in transgenic plants.** Considering that higher phytosiderophore secretion is correlated with higher Fe uptake in grasses and that rice secretes less phytosiderophore than other cereal plants [38], one logical approach towards enhancing Fe uptake in rice is to increase phytosiderophore production. Moreover, Fe uptake efficiency varies with the molecular nature of phytosiderophores [44]. Barley has the greatest capacity of phytosiderophore-mediated Fe uptake [45]; it was found that four barley genomic fragments, containing genes related to phytosiderophore biosynthesis, were independently transformed into rice plants [44].

IV. **Enhancement of root ferric reductase activity in transgenic plants.** Two ferric reductase genes were identified in the rice genome, OsFRO1 and OsFRO2, which are exclusively expressed in shoots [46]. Iron deficient transgenic plants had higher ferric reductase activity and higher iron uptake rates, than wild type plants [44].

V. **Supply of S.** It was found that high S supply increased Fe concentrations in the shoots of durum wheat plants, thus it was advantageous for plants grown under severe Fe limitation and could provide the opportunity to develop an agronomic practice to reduce the negative impact of Fe deficiency [47]; more specifically, Fe and S concentrations in the leaves of durum wheat plants were significantly correlated. This effect of S nutrition on Fe accumulation can be explained, according to the same authors, by an enhanced assimilation of S and its subsequent incorporation into methionine (since PS biosynthesis requires methionine) in order to sustain an increased production of phytosiderophores (PS); particularly, PS release rate increased with increasing S supply in wheat plants under Fe-limiting conditions [47]. This positive S nutritional effect on Fe uptake and accumulation should be also studied in detail in other plant species, which are grown under Fe starvation.

VI.   **Crop management practices.** There are agronomic (crop management) practices that can be adopted in order to increase Fe uptake by plants and to alleviate Fe chlorosis. One such practice could be the peanut mixed cropping with gramineous species on Fe chlorosis of peanut plants, grown in calcareous soils [11]. Apart from the mixed cropping, the assimilation in soil of the vegetal products from pruning (branches, leaves *etc.*) may contribute to the recycling of nutrients, thus to adequate supply of them (Fe in our case). For that purpose, vegetal products of pruning should not be destroyed or burned, but they should be assimilated in soils as composts, or organic fertilizers, in order to increase Fe availability and uptake.

VII. **Acid rain.** In areas with heavy atmospheric pollution acid rain is a phenomenon that may occur. Under these conditions soil pH may be decreased and Fe mobility is usually increased [1].

VIII. **The use of acid fertilizers.** When acid fertilizers are used (*e.g.*, NH_4^+ instead of NO_3^- fertilizers) rhizosphere is acidified and Fe mobility is increased, thus its' uptake by plants [1].

IX.   **Trunk injection.** Trunk injection under low pressure was found to be a good method to overcome Fe chlorosis in olive and peach trees [48]. This method could be possibly advantageous also for other tree species and for that reason it should be studied carefully in the near future.

X.   **Other practices.** Some simple and effective practices widely used by local farmers in China include Fe^{2+} root feeding (placing the cut roots in Fe^{2+}

fertilizer solution) and bag fertilization (placing the cut branches in a bag with Fe^{2+} fertilizer solution). These approaches make more Fe directly available for root uptake, thus correcting Fe deficiency in fruit crops; however, they are not feasible to treat entire fields, but are more suitable for labour-intensive small-scale fruit crop production in some areas with low labour costs [49].

Iron Uptake Mechanisms

There are two mechanisms adopted by plants in order to take up Fe from soil, strategy I, and strategy II. Strategy I is a complex Fe uptake mechanism developed by all plants, with the exception of *Poaceae* plants, which belong to strategy II [13]. This Fe uptake mechanism is based on the reduction of external Fe^{3+} to Fe^{2+} through the induction of Fe^{3+} chelate reductase enzyme, localized at the plasma membrane of the rhizodermal cells [50]. According to Molassiotis *et al.* (2005a), Fe(III) chelate reductase activity was higher in the (-Fe) treated roots, than in the (+Fe) treated ones of the peach rootstock GF-677 (*Prunus amygdalus* x *Prunus persica*). Furthermore, the induction of Fe(III) chelate reductase activity was accompanied by an increase in Fe, Zn and P concentration of explants [33]. The strategy I, developed by dicotyledonous plants to efficiently acquire Fe, involves the action of proton-extruding H^+-ATPases; these proteins are responsible for the solubilization of Fe through rhizosphere acidification [36]. In some strategy I species also, the release of organic compounds, such as phenolics, flavins, sugars and organic acids, could help in the solubilization of Fe-containing compounds [28].

In strategy II species, there is an increase in the synthesis and secretion of phytosiderophores (PS) to the rhizosphere [29]; then, the Fe-PS complexes are easily taken up by plants. In many grasses PS are secreted only during the morning after being accumulated, during the rest of the day and in the night [51]. It has been proposed that PS are synthesized in vesicles, most likely originating from the rough endoplasmic reticulum that appear in the cortex cells of Fe-deficient barley roots and accumulate in the epidermal cells at the cell periphery facing the rhizosphere just before sunrise [52].

Sensitive to Fe Chlorosis Plant Species

Many horticultural and agronomic crops, which are strategy I species, are sensitive to Fe deficiency; some of these species are: apple, grape, peach, *Citrus*, peanut *etc.* [49]. From strategy II species, rice and sorghum are among the most sensitive crops, compared to barley and wheat, due to their lower phytosiderophores excretion into the rhizosphere [53].

Prognosis and Identification of Fe Chlorosis

According to El-Jendoubi *et al.* (2012), it is possible to carry out the prognosis of Fe chlorosis in *Pyrus communis* trees by using early bud, flower and leaf mineral concentrations [54]. Also, bark analysis has been used for Fe deficiency prognosis in peach trees [55]. It has been proposed to identify Fe deficiency not only by leaf Fe concentration, but also by analyzing the concentrations of other nutrients. For example, the P/Fe ratio is considered to be a useful index to evaluate Fe chlorosis. This ratio increases when the chlorosis becomes severe [56] due to the increase of P uptake and decrease of Fe uptake. The same happens with the ratio K/Fe [57].

It was recently found that NO levels in roots was rapidly and continuously elevated when plants were transferred to a Fe-deficient growth medium [58], so its' measurement could be probably used as a tool for the detection of Fe chlorosis under conditions of Fe starvation.

Physiological Roles of Fe and Deficiency Symptoms

Since Fe is involved in chlorophyll formation and organic complexes are involved in the mechanisms of photosynthetic electron transfer [1], chlorophyll content and photosynthetic rates are usually decreased under Fe starvation; according to Molassiotis *et al.* (2006), Fe deficiency resulted in the reduction of the total chlorophyll content of peach rootstocks 'GF-677' and 'Cadaman'. Furthermore, Fe starvation caused a decline in photosynthetic rates, stomatal conductance and maximum quantum yield of PSII (F_v/F_m) [59]. In addition to the results of Molassiotis *et al.* (2006), it was found by Rombola *et al.* (2005) that Fe deficiency in sugar beet (*Beta vulagaris*) plants caused significant growth reductions, as well as reductions in chlorophyll content and net photosynthesis. Stomatal conductance and leaf transpiration rates, however, were not affected by Fe starvation [60], which was in contrast to the results of Molassiotis *et al.* (2006) for peach rootstocks 'GF-677' and 'Cadaman'. So, plant species is a crucial factor influencing the response to Fe chlorosis. The characteristic chlorotic appearance of whole plants suffering from Fe deficiency is usually the result of the decreased chlorophyll content of leaves, while the retard in plant growth and yields is usually associated with the effect of Fe starvation on several physiological processes and vital functions of plants [1], one of the most important of which is photosynthesis. Other vital functions of plants depending on Fe are enzymes' activity and nitrogen fixing. In Figs. (**1-4**) are presented leaf chlorosis symptoms in some famous cultivated species.

Fig. (1). Iron chlorosis in kiwi leaves (from Rhoades, http://www.gardeningknowhow.com/plant-problems/environmental/leaf-chlorosis-and-iron.htm) [65].

Iron is an important nutrient in N_2 fixing legume nodules. The demand for this micronutrient increases during the symbiosis establishment, where Fe is utilized for the synthesis of various containing proteins in both the plant and the bacteroid [61]. It was found that Fe is the primary micronutrient present in mitochondria [62]; mitochondria contain a large amount of metalloproteins that require Fe to carry out their function [63]. In fact, several enzymes belonging to both respiratory chain and to the tricarboxylic acid cycle are Fe-containing proteins; the morphology and ultrastructure of the mitochondria are also affected by Fe deficiency [64].

Fig. (2). Iron deficiency in corn (from Westfall and Bauder, http://www.ext.colostate.edu/pubs/crops/00545.html) [66].

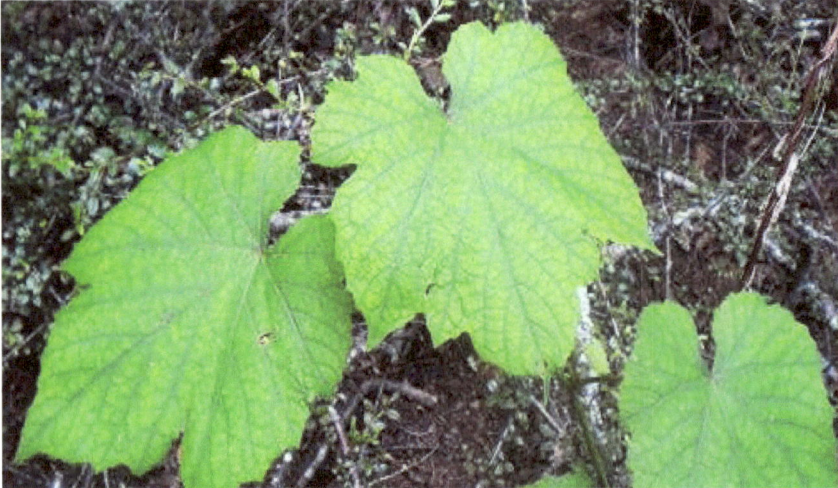

Fig. (3). Grapevine showing chlorosis of the leaves (from Day, http://www.todayshomeowner.com/how-to-treat-iron-deficiency-in-plants/) [67].

Fig. (4). Iron deficient leaves of *Beta vulgaris* grown in a controlled environment growth chamber in nutrient solution (A) and *Prunus persica*, grown in the field (from El-Jendoubi *et al.*, 2011) [57].

Mechanisms of Tolerance Adopted by Plants in Order to Face Fe Deficiency

A common consequence of most abiotic and biotic stresses (like Fe deficiency) is the increased production of reactive oxygen species (ROS), *i.e.*, superoxide (O_2^-), hydrogen hyperoxide (H_2O_2), hydroxyl radicals (OH^-) *etc.* The production of ROS leads to peroxidation of membrane lipids, base mutation, breakage of DNA strands and destruction of proteins [68]. According to Vigani *et al.* (2012), mitochondria generate several compounds as signals, such as ROS and organic acids [37]. More specifically, mitochondria produce ROS that not only cause damage to cellular components, but they are also involved in intracellular signalling: H_2O_2 is considered the most likely messenger [69]. Apart from H_2O_2, NO was recently found to be involved in the induction of Fe-deficient responses [37]; mitochondrion is probably the most affected compartment of the root cell under Fe deficiency, so it is clear that they have several signal types that might be involved in Fe deficiency signal transduction [37]. Under stress conditions (like under Fe deficiency) plant cells have an antioxidant defence mechanism against ROS constituting of: i) non-enzymatic low molecular weight antioxidants, such as ascorbate, a-tocopherol, glutathione and carotenoids and ii) enzymatic antioxidant system, according to the following reactions:

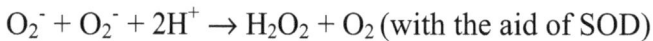

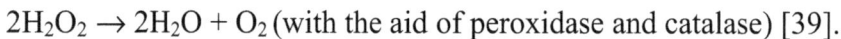

$$O_2 + e^- \rightarrow O_2^-$$

$$O_2^- + O_2^- + 2H^+ \rightarrow H_2O_2 + O_2 \text{ (with the aid of SOD)}$$

$$2H_2O_2 \rightarrow 2H_2O + O_2 \text{ (with the aid of peroxidase and catalase) [39].}$$

Great differences among genotypes exist concerning tolerance mechanisms to Fe starvation. Molassiotis *et al.* (2005b), who studied the antioxidant mechanisms in five *Prunus* rootstocks ('Peach seedling', 'Barrier', 'Cadaman', 'Saint Julien 655/2' and 'GF-677') differing in tolerance to Fe deficiency, found that the tolerance of these genotypes was associated with induction of an antioxidant defence mechanism (superoxide dismutase, peroxidase, catalase, isoforms of SOD, non-enzymatic antioxidants *etc.*) [70]. Genotypic differences in the mechanisms to Fe deficiency may be also related to the efficiency of the reductases activity or phytosiderophores secretion systems, in strategy I and strategy II plants, respectively [71].

In strategy I plant species, root system of tolerant plants exposed to Fe deficiency has an enhanced ability to induce H^+ extrusion. It was found that the tolerant to Fe starvation peach rootstock 'GF-677' exposed to (–Fe) treatment induced a strong H^+ extrusion, compared to the sensitive rootstock 'Cadaman' [59]. Iron efficient dicots are able to improve Fe uptake by enhancing root ferric chelate reductase and ATPase enzyme activities and also by increasing the release of reductants into the rhizosphere [72]. Indeed, between the susceptible to Fe deficiency common bean cultivar 'Coco blanc' and the tolerant 'Flamingo', the second showed greater tolerance due to a greater increase in the Fe-chelate reductase and H^+-ATPase activities in both roots and nodules, leading to a more efficient Fe supply to nodulating tissues [61]. More specifically, Fe deficiency induced the up-regulation of the expression of the ferric chelate reductase gene (*NtFRO1*), *via* a long-distance signal [73]. Wu *et al.* (2012) found that Fe deprivation in a portion of the root system of *Malus xiaojinensis* induced a dramatic increase in Fe(III) reductase activity and proton extrusion in the Fe-supplied portion, suggesting that Fe deficiency responses were mediated by systemic auxin (IAA) signalling. According to the same authors, IAA production in the shoot upon Fe deficiency was most likely triggered by a signal coming from the root system [74]. It was found that Fe-deficient sunflower roots produced higher levels of auxin, than the Fe-sufficient ones [72].

In strategy II plant species, genotypic differences in Fe deficiency stress tolerance are correlated with the quantities of PS that are released by roots in order to mobilize Fe (plants may uptake the complexes Fe-PS) under conditions of Fe starvation [75].

Some researchers found that plants grown under Fe deprivation modify the morphology of their root system (usually they increase the production of root hairs) in order to increase Fe uptake. For example, a more branched and rich in thin hairs root system is more capable of absorbing Fe, than a lateral one with poor development of thin absorptive hairs; it was found that between the olive cultivars 'Kothreiki' (a cultivar with richly branched root system), and 'Koroneiki' (a cultivar with a lateral, but poorly branched and with less development of absorptive hairs root system), the first one absorbed about 3 to 5 times more Fe (depending on the soil type where olive plants were grown), than the second one [76]. Zuchi *et al.* (2011) refer that the development of root hairs results in an increased cell volume of rhizodermis and may lead to an increased storage capacity for PS in roots. However, the same authors found that the presence of root hairs and their increased production in response to low Fe availability, while causing major modifications of root geometry did not necessarily lead neither to an effect on growth, nor on Fe uptake and accumulation in barley plants [77]. Apart from root architecture and morphology, another important factor that may influence Fe uptake by plants is the shoot to root biomass ratio [39].

Sources of Soil and Foliar Fe Fertilizers and their Efficiency in the Correction of Fe Chlorosis

There are two basic methods that Fe fertilizers can be applied to crops: either as soil or foliar application. Therefore, Fe fertilizers (applied either to soil, or to foliage), are supplied to many crops every year in order to control Fe deficiency. The quantity of Fe needed to overcome Fe chlorosis depends on crop [13]; for example, in peach trees 1-2 g./tree/year is needed [78]. Whereas the effect of Fe-fertilizers on Fe-chlorosis is dose dependent, different fertilizers could have different optimal doses [79]. Therefore, when comparing different fertilizers, the real effects could be masked by a dose effect: for instance, if an inefficient fertilizer is applied in a low quantity, no chlorosis correction will be observed [57]. Iron fertilizers are grouped into three classes: inorganic Fe compounds, synthetic Fe chelates, and natural Fe complexes [13]. Fertilizers based on inorganic Fe compounds include soluble substances, such as Fe salts (*e.g.*, $Fe_2(SO_4)7H_2O$) and insoluble compounds, such as Fe oxides-hydroxides and other cheap minerals and industrial by-products [80]. From organic (chelates) substances of Fe that can be used in order to alleviate Fe chlorosis, the use of Fe-EDTA and Fe-EDDHA should be distinguished. Recently, Sanchez-Alcala *et al.* (2012) proposed the use of synthetic siderite ($FeCO_3$) for the alleviation of Fe chlorosis in olive orchards, since it is an effective slow-release Fe, non-toxic and easy to prepare in the filed Fe fertilizer. Other very good solutions for the alleviation of Fe deficiency in olive orchards could be, according to the same authors, the use of vivianite (another effective slow-release Fe fertilizer), or vivianite plus humic acids, in addition to a solution of Fe chelate (EDDHA) [81].

The use of foliar fertilizers is very advantageous under alkaline soil conditions, where the solubility of Fe in soils is very low. Roosta and Mohsenian (2012) compared three Fe sources of foliar applications in pepper plants, in alkaline aquaponic solutions, and found that the best results (highest shoot Fe concentration and maximum quantum yield of PSII photochemistry) were obtained with the source of $FeSO_4$; similar results were found between Fe-EDTA and Fe-EDDHA, while the lowest Fe concentrations obtained in the control (untreated) plants [82].

According to Sorrenti *et al.* (2011), the soil-applied aqueous extract of *Amaranthus retroflexus* L. may solubilise Fe from calcareous soils; more specifically, it was found that the supply of *Amaranthus retroflexus* L. extract improved Fe nutrition of pear trees (*Pyrus communis* L.), particularly when enriched with $FeSO_4$, and increased shoot length and total plant biomass in controlled environment [83]. Recently, a new type of fertilizer, the glass matrix fertilizer (GMF), a by-product of ceramic industries, was applied in *Citrus* orchards suffering from Fe deficiency symptoms in calcareous and alkaline soils, with very promising results. More specifically, the mixture GMF and digested vine vinasse (DVV) reduced Fe chlorosis symptoms and increased Fe concentration, so it could be used as an 'environmental friendly' fertilizer allowing not only to reduce the use of chemical fertilizers, but also to re-use the industrial wastes and organic residues [84]. According to the same authors, the dried vine vinasse, constituted mainly by humo-similar organic compounds, is particularly able to complex mineral nutrients contained in GMF and making them available for the following root uptake [84].

El-Jendoubi *et al.* (2011) refer that using the same Fe-fertilizer under growth chamber and field conditions may give very different results; it was found in one of their experiments that the same commercial fertilizer was very effective in correcting Fe deficiency in peach grown in nutrient solution, in a growth chamber, but it was completely ineffective when applied in large quantity to Fe-deficient peach trees, grown on a calcareous soil in the field [57]. According to the same authors, the most appropriate way to assess the efficiency of Fe-fertilizers is to follow the evolution of leaf chlorophyll after Fe-fertilization. When using soil applications to fruit trees, once the Fe-fertilizer is applied, there is usually an approximately one-week lag phase, where no changes in leaf chlorophyll are observed. Then, leaf chlorophyll concentration begins to increase and the rapidity of this phase of the response may vary among different fertilizers during the first month after Fe-application; these differences likely reflect different Fe uptake and transport rates, which may depend on the specific product used [57].

REFERENCES

[1]   Kabata A, Pendias H. Trace elements in soils and plants. 3rd ed. CRC Press, USA, 2001.

[2]   Alifragis D. Soil: Genesis, Properties and Classification. Volume I. Aivazi Publications, Thessaloniki, Greece, 2008; pp. 493-496. (In Greek).

[3]   Bao T, Sun TH, Sun LN. Effect of cadmium on physiological responses of wheat and corn to iron deficiency. J Plant Nutr 2012; 35: 1937-1948.

[4]   Ghasemi-Fasaei R, Ronaghi A, Maftoun M, Karimian NA, Soltanpour PN. Iron-Manganese interaction in chickpea as affected by foliar and soil application of Iron in a calcareous soil. Com Soil Sci Plant Anal 2005; 36: 1717-1725.

[5]   Graham RD, Senadhira D, Beebe SE, Iglesias C. A strategy for breeding staple-food crops with high micronutrient density. Soil Sci Plant Nutr 1998; 43: 1153-1157.

[6]   WHO. Micronutrient deficiency: iron deficiency anaemia. Geneva: WHO 2007; Available from: http://www.who.int/en/

[7]   Morikawa CK, Saigusa M, Nakanishi H, Nishizawa NK, Mori S. Overcoming Fe deficiency in guava (*Psidium guajava* L.) by co-situs application of controlled release fertilizers. Soil Sci Plant Nutr 2006; 52: 754-759.

[8]   Alcantara E, Cordeiro AM, Barranco D. Selection of olive varieties for tolerance to iron chlorosis. J Plant Physiol 2003; 160: 1467-1472.

[9]   Donnini S, De Nizi P, Gabotti D, Tato L, Zocchi G. Adaptive strategies of *Parietaria diffusa* to calcareous habitat with limited iron availability. Plant Cell Environ 2012; 35: 1171-1184.

[10]   Zhu CF, Naqvi S, Gomez-Galera S, Pelacho AM, Teresa CT, Christou P. Transgenic strategies for the nutritional enhancement of plants. Trends Plant Sci 2007; 12: 1360-1385.

[11]   Zuo YM, Zhang FS. Effect of peanut mixed cropping with gramineous species on micronutrient concentrations and iron chlorosis of peanut plants grown in a calcareous soil. Plant Soil 2008; 306: 23-36.

[12]   Graham RD, Welch RM. Plant food micronutrient composition and human nutrition. Com Soil Sci Plant Anal 2000; 31: 1627-1640.

[13]   Grusak MA, Pearson JN, Marentes E. The physiology of micronutrient homeostasis in field crops. Field Crops Res 1999; 60: 41-56.

[14]   Arrieta L, Grez R. Solubilization of iron-containing minerals by soil microorganisms. Appl Environ Microbiol 1971; 22: 487-490.

[15]   Anonymous. Iron. Internet: http://hyperphysics.phy-astr.gsu.edu/hbase/pertab/fe.html

[16]   Lindsay WL, Norvell WA. Development of a DTPA soil test for zinc, iron, manganese and copper. Soil Sci Soc Am J 1978; 42: 6421-6428.

[17]   Irmak S, Surucu AK, Aydin S. The effects of Iron content of soils on the iron content of plants in the Cukurova region of Turkey. Inter J Soil Sci 2008; 3: 109-118.

[18]   Keramidas B. Fertility of Soils. Publications of the Aristotle University of Thessaloniki, Thessaloniki, Greece; 1997, pp. 73-74. (In Greek).

[19]   Fodor F, Kovacs K, Czech V, *et al*. Effect of short term iron citrate treatments at different pH values. J Plant Physiol 2012; 169: 1615-1622.

[20]   Abadia J, Vazquez S, Rellan-Alvarez R, *et al*. Towards a knowledge-based correction of iron chlorosis. Plant Physiol Biochem 2011; 49: 471-482.

[21]   Alcantara E, Cordeiro AM, Barranco D. Selection of olive varieties for tolerance to iron chlorosis. J Plant Physiol 2003; 160: 1467-1472.

[22]   Lindsay WL, Schwab AP. The chemistry of iron in soils and its availability to plants. J Plant Nutr 1982; 7: 821-840.

[23]   Fan X, Karim R, Chen X, *et al.* Growth and Iron uptake of lowland and aerobic rice genotypes under flooded and aerobic cultivation. Com Soil Sci Plant Anal 2012; 43: 1811-1822.

[24]   Fageria NK, Baligar VC. Response of common bean, upland rice, corn, wheat and soybean to soil fertility of an Oxisol. J Plant Nutr 1997; 10: 1279-1289.

[25]   El-Jaoual T, Cox DA. Manganese toxicity in plants. J Plant Nutr 1998; 21: 353-386.

[26]   Chatzistathis T. Investigation of the role of Mn on olive trees' mineral nutrition. Doctoral dissertation. Aristotle University of Thessaloniki, Greece. 2008. (In Greek).

[27]   Bridge T, Johnson D. Reduction of soluble iron and reductive dissolution of ferric iron-containing minerals by moderately thermophilic iron-oxidizing bacteria. Appl Environ Microbiol 1998; 64: 2181-2186.

[28]   Lopez-Millan AF, Morales F, Gogorcena Y, Abadia A, Abadia J. Metabolic responses in iron deficient tomato plants. J Plant Physiol 2009; 166: 375-384.

[29]   Lopez-Millan AF, Grusak MA, Abadia J. Carboxylate metabolism changes induced by Fe deficiency in barley, a strategy II plant species. J Plant Physiol 2012; 169: 1121-1124.

[30]   Shi RL, Hao HM, Fan XY, Md RK, Zhang FS, Zou CQ. Responses of aerobic rice (*Oryza sativa* L.) to iron deficiency. J Integr Agric 2012; 11: 938-945.

[31]   Tsednee M, Mak YW, Chen YR, Yeh KC. A sensitive LC-ESI-Q-TOF-MS method reveals novel phytosiderophores and phytosiderophore-iron complexes in barley. New Phytol 2012; 195: 951-961.

[32]   Schmidt W, Michalke W, Schikora A. Proton pumping by tomato roots. Effect of Fe deficiency and hormones on the activity and distribution of plasma membrane H$^+$-ATPase in rhizodermal cells. Plant Cell Environ 2003; 26: 361-370.

[33]   Molassiotis A, Therios I, Dimassi K, Diamantidis G, Chatzissavvidis C. Induction of Fe(III) chelate reductase activity by ethylene and salicylic acid in iron deficient peach rootstock explants. J Plant Nutr 2005a; 28: 669-682.

[34]   Romera FJ, Garcia MJ, Alcantara E, Perez-Vicente R. Latest findings about the interplay of auxin, ethylene and nitric oxide in the regulation of Fe deficiency responses by strategy I plants. Plant Signal Behav 2011; 6: 167-170.

[35]   Zhang XW, Dong YJ, Qiu XK, Hu GQ, Wang YH, Wang QH. Exogenous nitric oxide alleviates iron-deficiency chlorosis in peanut growing on calcareous soil. Plant Soil Env 2012; 58: 111-120.

[36]   Mlodzinska E. Alteration of plasma membrane H$^+$-ATPase in cucumber roots under different iron nutrition. Acta Physiol Plant 2012; 34: 2125-2133.

[37]   Vigani G, Chitto A, De Nisi P, Zocchi G. cDNA-AFLP analysis reveals a set of new genes differentially expressed in cucumber root apexes in response to iron deficiency. Biol Plant 2012; 56: 502-508.

[38]   Mori S, Nishizawa N, Hayashi H, Chino M, Yoshimur E, Ishihara J. Why are young rice plants highly susceptible to iron deficiency? Plant Soil 1991; 130: 143-156.

[39]   Marschner H. Mineral Nutrition of Higher Plants 1995; 2nd edition. Academic Press, London. pp. 324-333.

[40] Fuentes M, Ortuno MF, Perez-Sarmiento F, *et al*. Efficiency of a new strategy involving a new class of natural hetero-ligand iron (III) chelates (Fe(III)-NHL) to improve fruit tree growth in alkaline/calcareous soils. J Sci Food Agric 2012; 92: 3065-3071.

[41] Molassiotis AN, Dimassi K, Therios I, Diamantidis G. Fe-EDDHA promotes rooting of rootstock GF-677 (*Prunus amygdalus* x *Prunus persica*) explants *in vitro*. Biol Plant 2003; 47: 141-144.

[42] Singh G, Nathawat NS, Kishore N, *et al*. Differential translocation of ^{59}iron in iron sufficient and deficient sorghum plants. J Plant Nutr 2011; 34: 1723-1735.

[43] Prado RM, Alcantara-Vara E. Tolerance to Fe chlorosis in non-grafted quince seedlings and in pear grafted onto quince plants. J Soil Sci Plant Nutr 2011; 11: 119-128.

[44] Sperotto RA, Ricachenevsky FK, Waldow V, Fett JP. Iron biofortification in rice: it's a long way to the top. Plant Sci 2012; 190: 24-39.

[45] Romheld V, Marschner H. Genotypical differences among graminaceous species in release of phytosiderophores and uptake of iron phytosiderophores. Plant Soil 1990; 22: 147-153.

[46] Ishimaru Y, Suzuki M, Tsukamoto T, *et al*. Rice plants take up iron as a Fe^{3+}-phytosiderophore and as Fe^{2+}. Plant J 2006; 45: 335-346.

[47] Zuchi S, Cesco S, Astolfi S. High S supply improves Fe accumulation in durum wheat plants grown under Fe limitation. Env Exp Bot 2012; 77: 25-32.

[48] Fernandez-Escobar R, Barranco D, Benlloch M. Overcoming iron chlorosis in olive and peach trees using a low-pressure trunk-injection method. HortScience 1993; 28: 192-194.

[49] Zuo Y, Zhang F. Soil and crop management strategies to prevent iron deficiency in crops. Plant Soil 2011; 339: 83-95.

[50] Li L, Cheng X, Ling HQ. Isolation and characterization of Fe(III)-chelate reductase gene LeFRO1 in tomato. Plant Mol Biol 2004; 54: 125-136.

[51] Walter A, Pich A, Scholz G, Marschner H, Romheld V. Effects of iron nutritional status and time of day on concentrations of phytosiderophores and nicotianamine in different root and shoot zones of barley. J Plant Nutr 1995; 18: 1577-1593.

[52] Negishi T, Nakanishi H, Yazaki J, Kishimoto N, Fujii F, *et al*. cDNA microarray analysis of gene expression during Fe-deficiency stress in barley suggests that polar transport of vesicles is implicated in phytosiderophore secretion in Fe-deficient barley roots. Plant J 2002; 30: 83-94.

[53] Romheld V, Marschner H. Evidence for a specific uptake system for iron phytosiderophores in roots of grasses. Plant Physiol 1986; 80: 175-180.

[54] El-Jendoubi H, Igartua E, Abadia J, Abadia A. Prognosis of iron chlorosis in pear (*Pyrus communis* l.) and peach (*Prunus persica* l. batsch) trees using bud, flower and leaf mineral concentrations. Plant Soil 2012; 354: 121-139.

[55] Karagiannidis N, Thomidis T, Zakinthinos G, Tsipouridis C. Prognosis and correction of iron chlorosis in peach trees and relationship between iron concentration and Brown Rot. Sci Hortic 2008; 118: 212-217.

[56] Chouliaras V, Therios I, Molassiotis A, Diamantidis G. Iron chlorosis in grafted sweet orange (*Citrus sinensis* L.). Biol Plant 2004; 48: 141-144.

[57] El-Jendoubi H, Melgar JK, Alvarez-Fernandez A, Sanz M, Abadia A, Abadia J. Setting good practices to assess the efficiency of iron fertilizers. Plant Physiol Biochem 2011; 49: 483-488.

[58] Chen WW, Yang JL, Qin C, Jin CW, Mo JH, Ye T, *et al.* Nitric oxide acts downstream of auxin to trigger root ferric-chelate reductase activity in response to iron deficiency in *Arabidopsis*. Plant Physiol 2010; 154: 810-819.

[59] Molassiotis A, Tanou G, Diamantidis G, Patakas A, Therios I. Effects of 4-month Fe deficiency exposure on Fe reduction, mechanism, photosynthetic gas exchange, chlorophyll fluorescence and antioxidant defence in two peach rootstocks differing in Fe deficiency tolerance. J Plant Physiol 2006; 163: 176-185.

[60] Rombola AD, Gogorcena Y, Larbi A, Morales F, Baldi E, Marangoni B, Tagliavini M, Abadia J. Iron deficiency induced changes in carbon fixation and leaf elemental composition of sugar beet (*Beta vulgaris*) plants. Plant Soil 2005; 271: 39-45.

[61] Slatni T, Vigani G, Ben Salah I, *et al.* Metabolic changes of iron uptake in N_2-fixing common bean nodules during iron deficiency. Plant Sci 2011; 181: 151-158.

[62] Nouet C, Motte P, Hanikenne M. Chloroplastic and mitochondrial metal homeostasis. Trend Plant Sci 2011; 16: 395-404.

[63] Bertini I, Rosato A. From genes to metalloproteins: a bioinformatics approach. Wiley VCH Verlag 2007; pp. 2546-2555.

[64] Vigani G. Discovering the role of mitochondria in the iron deficiency-induced metabolic responses of plants 2012; 169: 1-11.

[65] Rhoades H. Leaf chlorosis and iron for plants: What does iron do for plants. Internet: http://www.gardeningknowhow.com/plant-problems/environmental/leaf-chlorosis-and-iron.htm

[66] Westfall DG, Bauder TA. Zinc and Iron deficiencies. Internet: http://www.ext.colostate.edu/pubs/crops/00545.html

[67] Day J. How to treat iron deficiency in plants. Internet: http://www.todayshomeowner.com/how-to-treat-iron-deficiency-in-plants/

[68] Mittler R. Oxidative stress, antioxidants and stress tolerance. Trends Plant Sci 2002; 7: 405-410.

[69] Moller IM, Rogowska-Wrzesinska A, Rao RSP. Protein carbonylation and metal-catalyzed protein oxidation in a cellular perspective. J Prot 2011; 74: 2228-2242.

[70] Molassiotis AN, Diamantidis GC, Therios IN, Tsirakoglou V, Dimassi KN. Oxidative stress, antioxidant activity and Fe(III)-chelate reductase activity of five *Prunus* rootstocks explants in response to Fe deficiency. Plant Growth Regul 2005b; 46: 69-78.

[71] Yehuda Z, Hadar Y, Chen Y. FeDFOB and FeEDDHA immobilized on sepharose gels as Fe sources to plants. Plant Soil 2012; 350: 379-391.

[72] Romheld V, Marschner H. Mobilization of iron in the rhizosphere of different plant species. Adv Plant Nutr 1986; 2: 155-204.

[73] Enomoto Y, Hodoshima H, Shimada H, Shoji K, Yoshihara T, Goto F. Long distance signals positively regulate the expression of iron uptake genes in tobacco roots. Planta 2007; 227: 81-89.

[74] Wu T, Zhang HT, Wang Y, *et al.* Induction of root Fe(III) reductase activity and proton extrusion by iron deficiency is mediated by auxin-based systemic signalling in *Malus xiaojinensis*. J Exp Bot 2012; 63: 859-870.

[75] Takagi S, Nomoto K, Takemoto S. Physiological aspect of mugineic acid, a possible phytosiderophore of gramineous plants. J Plant Nutr 1984; 7: 469-477.

[76]   Chatzistathis T, Therios I, Alifragis D. Differential uptake, distribution within tissues and utilization efficiency of Mn, Fe and Zn by olive cultivars 'Kothreiki' and 'Koroneiki'. HortSci 2009; 44: 1994-1999.

[77]   Zuchi S, Cesco S, Gottardi S, Pinton R, Romheld V, Astolfi S. The root hairless barley mutant *brb* used as model for assessment of role of root hairs in iron accumulation. Plant Physiol Biochem 2011; 49: 506-512.

[78]   Abadia J, Alvarez-Fernandez A, Rombola AD, Sanz M, Tagliavini M, Abadia A. Technologies for the diagnosis and remediation of Fe deficiency. Soil Sci Plant Nutr 2004; 50: 965-971.

[79]   Lucena JJ, Chaney RL. Response of cucumber plants to low doses of different synthetic iron chelates in hydroponics. J Plant Nutr 2007; 30: 795-809.

[80]   Shenker M, Chen Y. Increasing iron availability to crops: fertilizers, organo-fertilizers and biological approaches. Soil Sci Plant Nutr 2005; 51: 1-17.

[81]   Sanchez-Alcala I, Bellon F, Del Campillo MC, Barron V, Torrent J. Application of synthetic siderite ($FeCO_3$) to the soil is capable of alleviating iron chlorosis in olive trees. Sci Hortic 2012; 138: 17-23.

[82]   Roosta HR, Mohsenian Y. Effects of foliar spray of different Fe sources on pepper (*Capsicum annum* L.) plants in aquaponic system. Sci Hortic 2012; 146: 182-191.

[83]   Sorrenti G, Toselli M, Marangoni B. Effectiveness of *Amaranthus retroflexus* L. aqueous extract in preventing iron chlorosis of pear trees (*Pyrus communis* L.). Soil Sci Plant Nutr 2011; 57: 813-822.

[84]   Torrisi B, Trinchera A, Rea E, Allegra M, Roccuzzo G, Intrigliolo F. Effects of organo-mineral glass matrix based fertilizers on *Citrus* iron chlorosis. Eur J Agron 2013; 44: 32-37.

CHAPTER 4

Zinc Deficiency

Abstract: Zinc deficiency is one of the most important micronutrient deficiencies and many crops exhibit symptoms. There are many, soil and non-soil, factors influencing Zn solubility in soils and uptake by plants. Some of the most important factors are: pH, $CaCO_3$, organic matter, cation exchange capacity (C.E.C.), soil humidity and temperature, soil texture, parent material, the quantity of Fe oxides and hydroxides in soils, the interaction with other nutrients, the genotypic ability to absorb Zn (such as the differential exudation capacity among genotypes), the formation of mycorrhiza, the different management and agronomical practices adopted by local farmers during crop production *etc.* Zinc may be taken up by plants either as Zn^{2+} or as Zn soluble organic chelates; Zn-deoxymugineic acid (DMA) complexes are the most referred ones in bibliography and the most prefered for uptake by some plant species (barley). Generally, Zn seems to be a mobile nutrient, easily transferred between vegetal tissues; under Zn deficient conditions many plant species are able to mobilize limited, but crucial for plant growth, quantities of Zn from older-mature- leaves to younger ones.

Zinc is very closely involved in N, carbohydrate and lipid metabolism of plants, as well as in protein and RNA synthesis. Of great importance is also the role of Zn on root membrane permeability. Generally, Zn levels below 20 p.p.m. are considered as deficient, so inorganic fertilization or organic amendment is needed. The non-typical Zn deficiency symptoms are usually those related to depressed plant growth, since Zn starvation negatively influences IAA concentration; from the typical symptoms, it should be distinguished the formation of clusters or rosettes of small stiff leaves at the ends of the young shoots in fruit trees *etc.* Some of the tolerance mechanisms adopted by plants in order to survive under Zn starvation include the enhanced exudation ability by tolerant genotypes, the antioxidant mechanisms in order to detoxify reactive oxygen species (ROS), the formation of mycorrhiza, the enhanced mobilization and translocation ability of Zn (usually from older to younger leaves) in tolerant genotypes *etc.* When leaf Zn concentrations are below the critical limits, Zn soil or foliar applications are needed. For that purpose, many substances, such as $ZnSO_4$, or ZnEDTA may be used for the alleviation of Zn deficiency. However, some industrial by-products, varying from flue dust, reacted with sulphuric acid, to organic compounds, derived from the paper industry, may be also used for the correction of Zn starvation.

All the above mentioned topics concerning soil and non-soil factors influencing Zn solubility and uptake, the deficiency symptoms, the mechanisms of tolerance adopted by plants in order to face Zn starvation, as well as the methods of fertilizers' application and the substances used to alleviate Zn deficiency are within the purposes of this chapter and they are fully developed and discussed.

Keywords: Zn availability, Zn deficiency, ZnEDTA, Zn fertilizers, $ZnSO_4$, Zn tolerance, Zn uptake, Zn utilization efficiency.

INTRODUCTION

Zinc is an essential for plant growth mineral nutrient. About 30% of the world's soils are Zn-deficient [1] (Fig. **1**) and many crops exhibit symptoms of Zn deficiency. Furthermore, for human diet Zn starvation is the most serious micronutrient constraint, together with vitamin A deficiency. It is particularly widespread among children of the developing countries and represents a major cause of child death [2]. There are many soil factors influencing Zn availability and uptake by plants, such as pH, organic matter, $CaCO_3$, phosphoric ion content (P-induced Zn deficiency), cation exchange capacity (C.E.C.), soil moisture and temperature, the antagonistic or synergistic effect of other divalent ions on Zn absorption, the quantity of clay minerals and metal oxides *etc.* [3-6]. From the non-soil factors determining Zn solubility and uptake, they should be distinguished the over P-fertilization, the mobility of different organic and inorganic Zn fertilizers [7], the use of Zn humates [5], root exudations, like mugineic acid [8, 9], the formation of mycorrhiza, which is beneficial for Zn uptake [10], the agronomical management practices adopted by the local farmers in each region [11] *etc.*

So, under limited Zn availability it is very important to supply the necessary Zn quantities by applying Zn fertilizers, in order to avoid deficiency symptoms for crops and humans. According to Cakmak *et al.* (1999), the average grain Zn concentration of 54 wheat cultivars, grown on a Zn-deficient soil in central Anatolia, was 9 mg/kg dry weight (d.w.), while on the soils with normal supply of Zn, grain Zn concentration was 26 mg/kg d.w. [12]. The crucial Zn deficiency limit for normal plant growth depends on plant species, but it varies less among them than for other trace elements; generally it is around 10-15 mg/kg d.w. [3, 13]. The crucial Zn deficiency limit in soils, according to diethylenetriamine-pentaacetic acid (DTPA) method, is 0.5-0.6 mg/kg d.w. [14, 15]. Soils having concentrations below that limit are considered as Zn-deficient, thus they need Zn fertilization. Zinc deficiency may cause leaf chlorosis and depressed growth [16], or may exacerbate the adverse effects of short periods of heat stress on chloroplast function [17]. Zinc substances that can be used, either as soil, or foliar application, are $ZnSO_4$, ZnO, ZnEDTA *etc.* The use of Zn-enriched urea fertilizers seems to be a very promising strategy in the alleviation of Zn deficiency in crops and in the amelioration of Zn nutrition of humans and children in the developing countries, like India [2]. Foliar application is very advantageous in calcareous soils, where soil application is not effective in alleviating Zn starvation.

The purposes of this chapter are: i) to present and analyze the way by which different factors (soil, plant, as well as those provoked by human activities) influence Zn availability and uptake, ii) to describe the methods increasing Zn solubility in soils and alleviating Zn deficiency symptoms in plants, thus ameliorating Zn human nutrition, as well as iii) to provide all the important information concerning Zn fertilizers, their properties and solubility, and iv) to highlight the mechanisms adopted by plants in order to face Zn starvation.

Zn in Soils

Mankeze *et al.* (2012) [18] found that EDTA extractable soil Zn ranged from 0.50 to 2.43 mg/kg dry weight of soil. Phattarakul *et al.* (2012) [19] studied the soil and rice Zn status in China, India, Lao PDR, Thailand, and Turkey and found that in a pH range from 4.8 to 8.8 the DTPA extractable Zn concentrations varied from 0.5 to 6.5 mg/kg soil. Generally, DTPA extractable Zn concentrations below 0.5 mg/kg soil are considered deficient [14]. Critical soil Zn levels for sorghum (cvs. 'PARC-SS-1' and 'Potohar 4-8') were 3.1-3.4 mg/kg, according to DTPA and 3.5-3.7 mg/kg, according to the extractant solution AB-DTPA. These critical quantities were much higher (7.2-8.0 mg/dm^3) when Mehlich-3 was used for extraction [20]. There are two different mechanisms of Zn adsorption: one in acid media related to cation exchange sites, and the other in alkaline media that is highly influenced by organic ligands. The adsorption of Zn^{2+} can be reduced at lower pH (<7) by competing ions and this phenomenon may result in easy mobilization and leaching of Zn from light acid soils. At higher pH values, while an increase of organic compounds in soil solution becomes more evident, Zn-organic complexes may also account for the solubility of Zn [21]. Free Zn^{2+} ions may adsorb onto organic matter in a region of low pH and may precipitate as franklinite or other minerals at high pH [22]. Generally, the solubility and availability of Zn in soils is basically the result of the interaction between pH, soil organic matter, cation exchange capacity (C.E.C.) and texture. Factors decreasing solubility and mobility of Zn in soils stimulate adsorption onto soil constituents, such as clay minerals and metal oxides [23]. The role that each one of these soil factors plays on Zn solubility and uptake by plants is fully analyzed below.

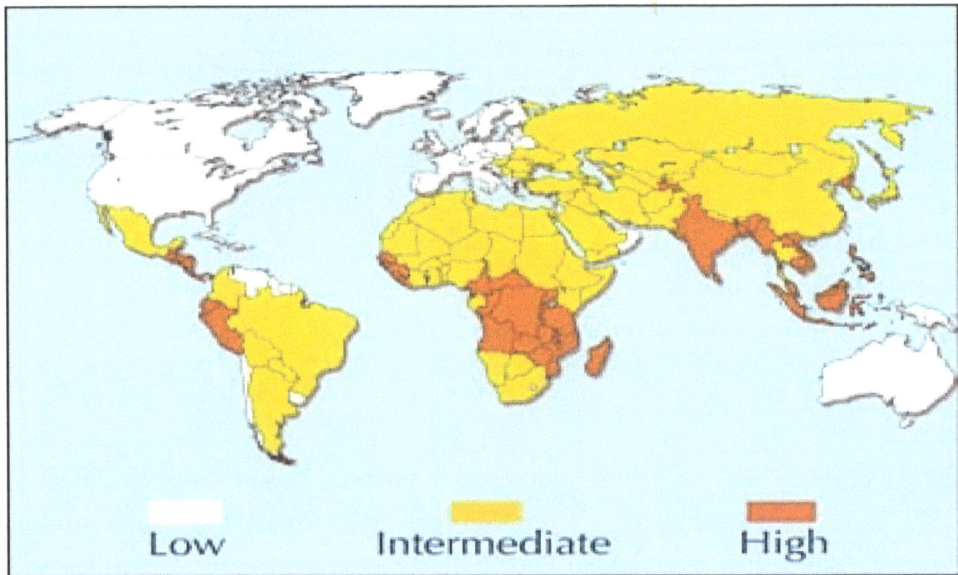

Fig. (1). Distribution of Zn deficient areas around the world (http://www.izincg.org/) [24].

Factors Influencing Zn Availability in Soil and Uptake by Plants

The factors influencing Zn solubility and uptake by plants may be divided in three categories: soil factors, plant factors and factors influenced by human activities. There are enough soil factors increasing or decreasing Zn availability and uptake: pH, $CaCO_3$, soil humidity, organic matter, parent material, the presence and concentration of other nutrients, acting synergistically or antagonistically on Zn uptake *etc.* Mean total Zn contents in surface soils of different countries and of the U.S. vary from 17 to 125 p.p.m. Mean Zn for worldwide soils may be calculated as 64 p.p.m. [21].

Soil Factors

I.   **pH and $CaCO_3$.** Soil acidity influences Zn solubility; Zn availability is highly dependent on soil pH and is very low at high soil pH. It is particularly low when $CaCO_3$ is present because of specific adsorption of

Zn^{2+} to and occlusion by carbonates [13]. According to Mengel and Kirkby (2001), adsorption and occlusion of Zn by carbonates are the major causes of poor Zn availability and appearance of Zn deficiency symptoms on calcareous soils [13]. In contrast to that, the adsorption of Zn^{2+} can be reduced at lower pH values (pH<7) by competing ions and this phenomenon may result in easy mobilization and leaching of Zn from light acid soils (soils with low cation exchange capacity-C.E.C. values) [21]. Zinc is present with the form Zn^{2+} when pH is lower than 7.7, whilst in greater values it is present in the form of $Zn(OH)_2$ [25]; however, Zn deficiency in most soil types takes place when pH is higher than 6 [26].

II.   **Salinity.** Alkaline soils have high amounts of $CaCO_3$, which is the cause for the decreased Zn availability. Wheat plants grown on such soils suffer from Zn deficiency, although great genotypic differences concerning Zn uptake exist [27].

III.   **Soil humidity.** Dry-lands in the semi-arid regions of India suffering from water shortage are more susceptible to widespread deficiencies of S, B and Zn, than other soil types with sufficient soil water content [28]. Under these adverse soil conditions Zn solubility is significantly decreased and it is not available for plants. The availability of Zn is also reduced by flooding due to the prevalence of anaerobic conditions (Zn^{2+} uptake is a metabolic process requiring enough oxygen) or due to leaching, especially in light sandy soils.

IV.   **Soil temperature.** Rice is prone to Zn deficiency under cold weather conditions and low soil temperatures [13]. Low soil temperatures may be also observed in very humid and flooded soils.

V.   **Soil conditions restricting root growth.** Factors limiting the rate of diffusion of Zn to plant roots may also reduce Zn availability. This is probably the most important reason why Zn deficiency occurs on compacted soils, or where root growth is restricted [13].

VI.   **Soil texture.** Soil texture greatly influences Zn availability: The lowest values were found in light mineral and light organic soils. Geometric means in mineral soils of Poland significantly differ for various textural groups: for example, the mean value for sands is 33 p.p.m., while for light and heavy loams the mean values are 52 p.p.m. and 80 p.p.m., respectively [21]. The lower Zn concentrations in light soils are owed to the enhanced leaching, especially in soils with high soil humidity and low organic matter content. In clayey soils, Zn^{2+} is usually adsorbed from clay colloids less than other nutrients; this is why Zn is considered to be more labile nutrient, compared to others [25].

VII. **Organic matter.** Zinc interacts with soil organic matter, and both soluble and insoluble organic complexes are formed. On average, about 60% of the total soluble Zn in soils occurs as soluble organic complexes; soluble Zn organic complexes are mainly associated with amino, organic and fulvic acids, while insoluble organic complexes are derived from humic acids [13]. The solubility or insolubility of Zn organic complexes depends not only on the kind of organic substances, but also on soil pH; in some cases (alkaline soils), soluble organic complexes greatly contribute to Zn nutritional needs of plants. According to Kabata-Pendias (2001), Zn-organic complexes may account for Zn solubility at higher pH values [21]. In any case, Zn deficiency more often appears in soils with low organic matter content, than in soils having sufficient organic C [25].

VIII. **Parent material and Zn-containing minerals.** The level of Zn in soils is very much related to the parent material and to the quantity of Zn-containing minerals. Soils originating from basic igneous rocks are high in Zn. In contrast, soils derived from more siliceous parent materials are particularly low in Zn [13]. Zinc is contained in many oxides, such as (ZnO), sulfides (ZnS), sphalerite-(ZnFe)S, carbonate salts ($ZnCO_3$) *etc.* The two Zn-silicates $ZnSiO_3$ and Zn_2SiO_4 (willemite) also occur in some soils [13, 21, 25]. Zinc soil content is referred to vary from 1.0 to 300 p.p.m. However, from this quantity only a percentage of less than 5% is in exchangeable form [25].

IX.   **The kind of clay minerals, the presence and quantity of Fe oxides and hydroxides** *etc.* When montmorillonite is the clay mineral that is basically present in soil, Zn may enter its layer lattice structures and become very immobile and not available for plants [21]. So, in montmorillonitic soils it is more possible to face Zn deficiency, than for example in the caolinitic ones. According to Abd-Elfattah and Wada (1981), the highest selective adsorption of Zn was found in Fe oxides, halloysite and allophane [29].

X.   **Interaction with other nutrients.** According to Cakmak and Hoffland (2012), care should be taken about interactions of Zn with other macronutrients, such as N within plants or P in soils, that could interfere with uptake, root-shoot transport and seed deposition of Zn [30]. It was found that, increasing N concentrations from low to medium levels, in a hydroponic experiment with *Triticum durum* (cv. Balcali 2000) plants, the total plant Zn content increased by 33% at anthesis and up to 60% at the maturity stage. However, high N application rates negatively affected Zn concentrations in tissues due to the dilution effect (it stimulated vegetative growth and dry matter production) [31]. Particularly in soils receiving rich P fertilization, insoluble Zn phosphate substances are formed, so Zn availability is reduced. From that point of view, excess P fertilizations should be avoided. According to Kabata and Pendias (2001), the specific mechanisms of Zn-P interaction are not yet known; however, the ratio Zn/P in plant tissues should be taken into consideration when discussing about this interaction. For example, it was found that the optimum P/Zn ratio for corn plants is 100. Zn-N interaction is mostly a secondary 'dilution' effect due to the increase of biomass because of the heavy N treatment [21]. Zinc uptake by plants may be also reduced due to competition by other cations (*e.g.*, Ca, Mg, Na) at the root surface [32]. The negative influence of excess Ca on Zn uptake may be direct (due to competition in plant uptake by divalent ion, *i.e.*, Ca^{2+}/Zn^{2+}), or indirect *via* the elevation of soil pH and decrease of Zn solubility. Graham *et al.* (1987) refer that inadequate Zn content leads to increased uptake of B and S by barley [33]. It was also found that Cu^{2+} strongly inhibited Zn^{2+} uptake; it seems possible that these two ions compete for the same uptake system. Finally, depressing effects by

Cd, Ba and Sr on Zn uptake have been referred [13]. Especially for the effect of Cd on Zn uptake, it is referred that this interaction appears- according to the findings of many researchers to be somewhat controversial, since there are reports of both antagonism and synergism between these nutrients in the uptake-transport processes [21].

Plant Factors

I.   **Phytosiderophore exudation.** Daneshbakhsh *et al.* (2013) found that the wheat Zn efficient genotypes 'Cross' and 'Rushan' released higher amounts of phytosiderophores (in order to increase Zn uptake), compared to the Zn inefficient 'Kavir' [34].

II. **Mycorrhiza.** Because of the low mobility of Zn in many soils mycorrhiza can substantially increase its uptake, like that of phosphate, by the host plant [10]. Apart from the direct influence of mycorrhiza on the increase of Zn availability and uptake, an indirect effect of it on Zn absorption may be explained *via* its' influence on root geometry and growth. It was found that arbuscular mycorrhiza fungus (AMF) affected root system morphology and nutrient uptake [35]. We found in one of our recent experiments that the differential AMF colonization was probably the reason for the differential root morphology pattern of three Greek olive cultivars and, thus, for the significant differences among genotypes in leaf Zn concentration and total Zn content [36].

Factors Influenced by Human Activities

I.   **The addition of wastes and sewage sludges in soils.** Occasionally, high levels of Zn may occur in soils affected by wastes and sludges [13]. This practice may be very useful in order to increase Zn availability in soils having very low Zn concentrations. According to Kabata and Pendias (2001), the addition of sewage sludges modifies the distribution of pattern of Zn, increasing significantly the easily soluble and exchangeable fraction of Zn [21].

II.   **Atmospheric input of Zn in polluted areas.** In polluted areas atmospheric input exceeds its output due to leaching and biomass harvesting. Only in non-polluted forest regions of Sweden the discharge of Zn by water flux reported to be higher than its atmospheric input [21].

III.   **Farming, management and agronomical practices used during crop production.** In cropping areas with intensified farming, additional management practices (Zn fertilization) are often required to avoid Zn depletion of soils and to sustain crop productivity and nutritional quality of the harvested product. Foliar application of Zn fertilizers represents a short-term and highly effective strategy to increase grain Zn concentration. Finally, breeding for high grain Zn is also an important solution and discussed as a cost-effective and long-term solution [30]. However, it still needs to be evaluated across locations with different management practices, since grain Zn concentrations may be significantly affected by locations and management [37]. In fruit crop areas where no Zn-fertilizers are applied after pruning and fruit collection (removal of Zn from agro-ecosystems), plants next season may suffer from Zn deficiency and fruits (like apples) will become small, without commercial value, since Zn soluble sources in soil are not sufficient to cover their nutritional needs. Zinc deficiencies most frequently result from management practices used during crop production (*e.g.*, overliming, P fertilization, organic matter amendment) [21]. More specifically, liming, a practice that is used to decrease metal solubility in areas where crops suffer from heavy metal toxicity, when applied in exaggerate amounts may lead to Zn deficiency. In one of our experiments, conducted to decrease Cu and Mn toxicity in old mine rural areas, we limed *Populus* sp., *Eucalyptus* sp., *Robinia pseudoacacia* and *Juglans regia* plantations, and we found that together with the significant reductions of toxic Mn and Cu concentrations in leaves, Zn leaf concentrations were also reduced from 67% in *Populus* sp. plantations up to 3 times in *Eucalyptus* sp. However, leaf Zn concentrations after liming were above the critical limit of Zn deficiency (Chatzistathis, unpublished data). In crops receiving over P-fertilization, insoluble Zn phosphate substances are formed, which are not available to plants. Finally, organic matter amendment is a good management practice used to ameliorate soil physical properties and to enhance nutrient availability. The different kind of manures used as organic fertilizers, are valuable sources of Zn, since they were found to contain 15-90 p.p.m. Zn; so, chicken manure usually contains higher nutrient content than those of cows, horses and sheep [16]. Other management practices, such as rotation, or the amendment with woodland leaf litter should not be omitted to be referred as important agronomic factors influencing Zn availability in soils and uptake by plants; legume-maize rotation was applied in the Zn deficient soils of eastern Zimbabwe with very good results (extractable soil Zn-according to EDTA- concentration was within the range 1.6-2.1 mg/kg soil), while the soils amended with woodland leaf litter reached the EDTA extractable concentration of 2.4 mg/kg soil Zn, compared to those of the fields not received Zn fertilization, which ranged from 0.5 to 0.7 mg/kg soil. However, the best management practice was found to be the combined use of cattle manure with inorganic fertilizers, which increased about 6 times the net amount of Zn taken up by maize plants [18].

Zn Uptake, Mobilization and Distribution in Plants

Zinc may be taken up by plants either as Zn^{2+}, or as soluble organic chelates. According to Kabata and Pendias (2001), who made a review of many scientific papers on Zn uptake, disagreement exists in the literature whether Zn uptake is a metabolically active or a passive process; however, several controversial results strongly suggest that Zn uptake is mostly metabolically controlled. The same authors also refer that apart from free Zn^{2+} existing inside plant organism, Zn may be also bound to soluble low-molecular weight proteins or light organic compounds in xylem fluids [21]. In barley for example, the uptake of Zn-deoxymugineic acid (DMA) complexes is preferred to Zn^{2+} through roots [38]. In contrast to that, rice plants take up less Zn-DMA complexes compared to Zn^{2+} [39]. The biosynthesis of the mugineic acid family phytosiderophores (MAs) and their corresponding genes have been characterized, and it was found that MAs are synthesized from methionine. In the cases of Zn deficiency, DMA synthesis is induced in barley shoots, while MAs synthesis and secretion by roots are induced under both Zn and Fe deficiency [40].

Generally, Zn seems to be a mobile nutrient, easily transferred between vegetal tissues; under Zn deficient conditions many plant species are able to mobilize limited, but crucial for plant growth, quantities of Zn from older-mature- leaves to younger ones, or other generative organs. Other valuable 'pool' of Zn which may be mobilized to cover nutritional needs of plants under Zn limitation is the 'storing capacity' of the root system, since in most cases root Zn concentrations are much greater to those determined in leaves and shoots, especially in tree species. That phenomenon of enhanced translocation from root system to shoots gets more important under luxury levels of Zn in soils [21]. Despite the fact that the uptake of Zn^{2+} is preferred to the uptake of Zn-DMA complexes by rice plants, translocation from root system to shoot is greater as Zn-DMA complexes, than as Zn^{2+} [40]. Translocation and remobilization of Zn was found to be critical for grain Zn accumulation when its' availability is restricted during grain filling [31]. Palmgren *et al.* (2008) suggested that in wheat and barley Zn remobilization contributes to grain Zn accumulation more than continued uptake [41].

It was recently found that several members of the Zn-regulated transporters (like ZIP family genes) have been characterized and shown to be involved in Zn uptake and transport in rice. Particularly, OsZIP1 and OsZIP3 genes seem important for Zn uptake from soil, while OsZIP4, OsZIP5 and OsZIP8 for root to shoot translocation [42]. According to Assuncao *et al.* (2010), 15 genes (8 out of which are members of the ZIP family and whose their exact role is largely unknown) are involved in the Zn uptake and mineral nutrition of *Arabidopsis* plants [43].

Roles of Zn in Plants

Zinc is completely necessary for the synthesis of tryptophan (former substance of auxin-IAA) [16]. So, under Zn starvation plant growth is suppressed due to low auxin concentration. Under Zn deficiency conditions, decreases in shoot dry matter production varied from 16% (in rye cultivars) to 47% (in durum wheat cultivars) [44]. Apart from the role of Zn in the synthesis of IAA, it is also constituent of other enzymes, such as metalloenzymes and dehydrogenases (dehydrogenase of glutamic acid, dehydrogenase of L-galactic acid *etc.*) [16]. Furthermore, Zn is very closely involved in N metabolism of plants, protein (under Zn starvation protein levels are markedly reduced and amino acids and amides are accumulated) and RNA synthesis (RNA polymerase contains Zn and when there is lack of Zn this enzyme is inactivated and RNA synthesis is impaired), as well as in carbohydrate and lipid metabolism [42]. Since the enzyme catalyzes the transcription of DNA and transfers the information of DNA to RNA, it is evident that the universal importance of Zn on the multiplication of genetic information and the transfer of genetic information on protein synthesis is very high [13].

Under high salinity conditions, it was found that adequate Zn nutrition counteracted the detrimental effect of high NaCl level on the growth of three wheat genotypes [45]. Recently, a very important role of Zn on root membrane permeability under high salinity, or any other kind of stress, was found to exist: Root membrane permeability is highly related to the production of reactive oxygen species (ROS). The production of ROS and superoxide radicals may disturb root membrane permeability and lead to ion leakage. For wheat cultivar 'Cross', supplied with adequate Zn, lower root membrane permeability and higher –SH group content were found at high salinity levels. More specifically, at moderate to high salinity levels Zn-deficient roots of three wheat genotypes leaked significantly higher amounts of Fe and K, than the Zn-sufficient roots [45]. In addition to the above, Cakmak (2000) refers that Zn plays a central role in the detoxification of ROS in plant cells [46].

It is also referred by Mengel and Kirkby (2001) that the sunscald, which is characterized by the destruction of photosystems and membranes by oxygen radicals and leading to a severe loss of fruit and vegetable quality, frequently occurs in countries with hot climates and high light intensities. Since plant species which are resistant to sunscald are related to have high SOD concentrations in their tissues, it was found that plants well supplied with Zn are protected in this respect [13]. As Zn affects membrane integrity and activity of superoxide dismutase (SOD), the effect of Zn on heat stress was investigated by Peck and McDonald (2010) and it was found that low Zn levels can exacerbate the adverse effects of short periods of heat stress on bread wheat plants and chloroplast function (chlorophyll fluorescence and chlorophyll content were reduced and chloroplast ultrastructure was disrupted by heat stress); more specifically, when both stresses (heat stress and Zn deficiency) occurred together, the severity of damage increased markedly [17].

It was recently found that priming rice seed with Zn can improve germination and enhance seedling vigor. However, this happened until the concentration of 5mM $ZnSO_4$ in nutrient solution; higher Zn concentrations (10 and 25mM) depressed seedling vigor [47]. High seed Zn can also help to alleviate pathogen infection under Zn deficient conditions, as Zn plays a fundamental role in protecting and maintaining structural stability of cell membranes and thus decreasing root exudation of various organic compounds that may stimulate pathogenic infection [46].

Zinc Concentrations in Plants and Deficiency Symptoms

The level of Zn in Zn-deficient plants is low and it is usually within the range 0-15 mg/kg d.w. According to Kabata and Pendias (2001), the deficiency content of Zn in plants has been established at 10 to 20 p.p.m.. These values may vary considerably because Zn deficiency reflects both the requirements of each genotype and effects of the interactions of Zn with other elements within plant tissues [21]. In strawberry plants, levels below 20 p.p.m. warrant the addition of Zn; the ideal Zn concentrations in the leaf tissue of strawberries is about 35 p.p.m. (levels in the range 20-50 p.p.m. are normal) [48]. According to Ozkutlu *et al.* (2006), the shoot concentrations of wheat and soybean plants suffering from Zn deficiency were 6 and 8 mg/kg d.w., respectively [49]. It was found in a survey, conducted in Poland, that the mean Zn content of over 5000 samples of potato tubers (grown in Poland) was 20 p.p.m., within the range varying from 1.2 to 150 p.p.m. Finally, a survey for food composition in the US gave the following values of Zn in some categories of plant food: i) vegetables (range from 0.7 to 8.0 p.p.m.), ii) fruits (range from 0.4 to 3.0 p.p.m.), iii) cereals (range from 0.7 to 32.5 p.p.m.), iv) nuts (range from 5.0 to 42.3 p.p.m.) [21].

Zinc deficiency symptoms may be categorised into typical and non-typical, their variety of expression and severity of which depends also on plant species. The non-typical symptoms are usually those related to depressed plant growth, since Zn starvation negatively influences IAA concentration [16]. A variety of visual typical Zn deficiency symptoms has been referred, such as whitish-brown necrotic patches on leaf blades of wheat plants [44], interveinal chlorosis in *Juglans regia* [16], chlorotic bands form on either side of the midrib of the leaf in the monocots [13] *etc.* General typical symptoms of Zn deficiency for many plant species, especially for fruit trees, are the formation of small leaves and fruits (without commercial value), as often happens with apples [16]. Unevenly distributed clusters or rosettes of small stiff leaves are formed at the ends of the young shoots. Frequently, the die off and the leaves fall prematurely. In apple trees the disease occurs in the early part of the year and is known as rosette or little-leaf. In *Citrus* trees the main veins remain green, as if the leaves were recovering from Fe deficiency (Fig. **2**). In vegetable crops, symptoms of Zn deficiency are more

Fig. (2). Zinc deficiency symptoms in orange leaves [51].

species-related (Figs. **4**, **6**), than deficiency symptoms of the other nutrients [13]. In strawberry plants, green margins or halos along the edges of young leaves may be observed; the blades of these leaves will become thinner and narrower; then the leaves will turn yellow and the number of fruits per plant will decrease, even if the average size of each fruit remains normal (Fig. **3**). In cotton plants the typical Zn deficiency symptoms are exhibited approximately after 3 weeks from sowing (Fig. **5**); some characteristic symptoms in cotton plants are those related to upward cupping of the leaf edges, flowers fall, maturity delay and deterioration of the fiber quality. In soybean plants, chlorosis and brown patches on young leaves have been referred, while in wheat plants necrotic spots on middle-aged leaves were found [49].

Fig. (3). Zinc deficiency symptoms in strawberry plants [48].

However, the sensitivity of genotypes of the same species may be completely different; according to Cakmak *et al.* (1997a), who studied the sensitivity to Zn deficiency of two bread and two durum genotypes, necrotic patches on leaf blades appeared rapidly and were severe in bread wheat cultivar 'BDME-10', and particularly in durum cultivars 'Kunduru-1149' and 'Kiziltan-91', while the bread genotype 'Bejostaja-1' was much less affected by Zn deficiency [44]. In another study of Cakmak *et al.* it was found that the decreases in shoot dry matter production due to Zn deficiency were absent in rye (*Secale cereale* L., cv. 'Aslim') and on average 5% in Triticale, 34% in bread wheat cultivars and 70% in durum wheat cultivars [50]. The differential genotypic sensitivity to Zn deficiency may be ascribed to many tolerance mechanisms adopted by plants in order to overcome Zn deficiency. These mechanisms are analyzed in detail below.

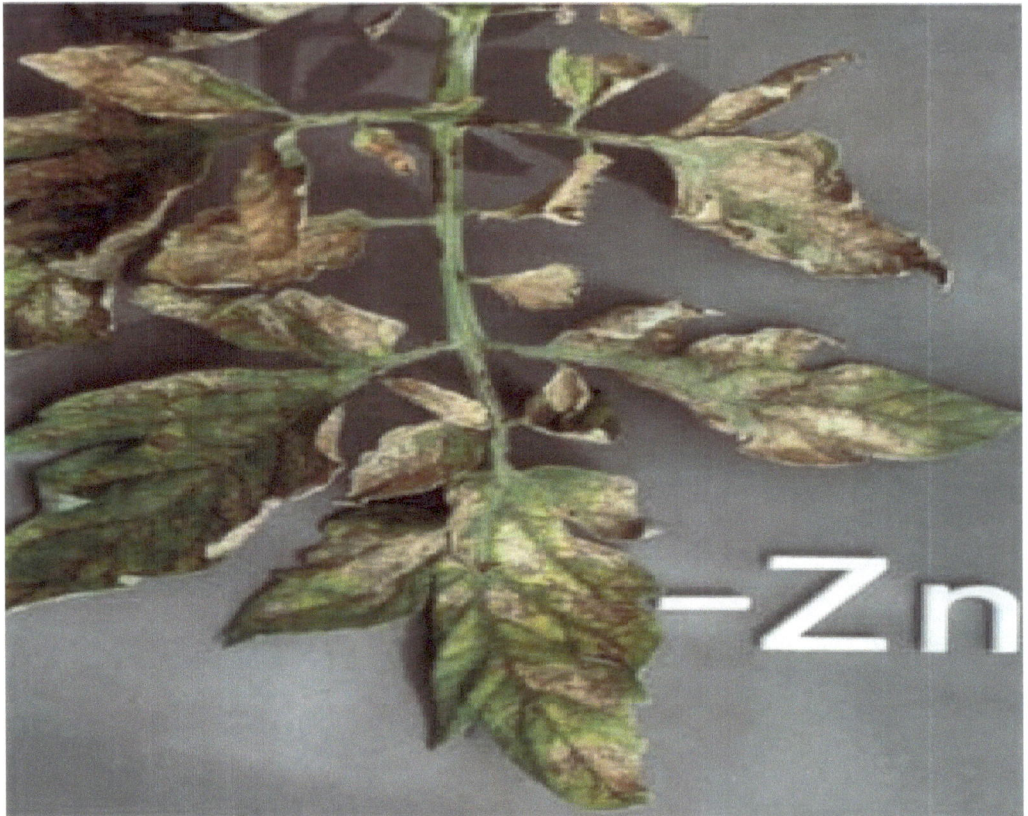

Fig. (4). Zinc deficiency symptoms in *Solanum lycopersicon* plants [52].

Zn Efficiency Among Genotypes

Zinc deficiency symptoms are not always related to low Zn concentrations in plant tissues, but rather to low Zn uptake and utilization efficiency by plants. According to Cakmak *et al.* (1997b), the very distinct differences among and within the cereal species (rye, triticale, bread and durum wheat) in susceptibility

Fig. (5). Zinc deficiency in cotton plants [53].

Fig. (6). Zinc deficiency symptom in lettuce [54].

to Zn deficiency were closely related with the total Zn content, but not with the Zn concentrations in shoot dry matter (the total amount of Zn was approximately 6 times higher in rye, than in the durum wheat). These results were ascribed, according to the same authors, to the exceptionally high Zn efficiency of rye, while the genotypic expression of Zn efficiency were possibly related to a much greater capacity of efficient genotypes to acquire Zn from soil, compared to inefficient ones (such as those of bread and durum wheat in the study of Cakmak *et al.*) [50]. According to Rengel and Graham (1996), the differential Zn efficiency of wheat genotypes is at least partly due to a greater ability of efficient genotypes to accumulate Zn [55].

Rice genotypes greatly differ in Zn use efficiency and this characteristic is correlated to different grain Zn contents; rice grain Zn concentrations was found to range from 15.9 to 58.4 mg/kg [56], suggesting ample variation for this trait that might be exploited through conventional breeding [40]. Rice seed has a very low Zn content compared with other cereals, such as wheat and barley, which can result in adverse effects on seedling growth and development; Chanakan *et al.* (2012) found that germination and seedling growth of rice cultivar TDK7 was still limited in seed containing 23 mg Zn/kg [47]. It was found by Wu *et al.* (2010) that Zn concentration in rice grains is closely associated with the extent of Zn retranslocation from source tissues to grains, so it is clear that genotypes having great ability to mobilize and retranslocate Zn from source tissues to grain are highly Zn efficient, compared to those having limited Zn mobilization ability, and should be selected and preferred for cultivation in Zn deficient soils [57].

Mechanisms of Tolerance Adopted by Plants to Overcome Zn Deficiency

There are mechanisms adopted by plants in order to overcome Zn deficiency in calcareous, alkaline and any other kind of Zn-deficient soils. Some of these mechanisms leading to the improved Zn uptake under Zn deficient conditions are the differential phytosiderophore exudation capacity among genotypes, the root system architecture and modification, the formation of mycorrhiza *etc.* All these mechanisms are fully analyzed below.

I.   **Phytosiderophore release.** In Zn deficient soils, Zn uptake may be enhanced by the exudation of low-molecular weight compounds, like malate and mugineic acid family phytosiderophores [38, 58]. According to Daneshbakhsh *et al.* (2013), the highest amount of phytosiderophores by the roots of the wheat genotype 'Kavir', grown hydroponically, released at the highest salinity level (120 mM NaCl), while for cultivars 'Cross' and 'Rushman' no difference was found in phytosiderophore exudation between the treatments 60 and 120 mM NaCl [34]. It was found to exist a good relationship between root exudation of phytosiderophores and differential tolerance to Zn deficiency when durum (sensitive) and bread wheat (tolerant to Zn deficiency) were tested [8, 59].

II. **Differential antioxidant capacity.** Similarly to other stresses, Zn deficiency induces oxidative stress in plants by affecting both generation and detoxification of O_2 free radicals. The production of reactive oxygen species (ROS) under any kind of stress is a very serious problem and concern about the functional permeability of membranes. For that reason, under Zn deficiency stress supply of Zn is required in order to detoxify ROS, including superoxide radical and H_2O_2. It is referred that plants may become more sensitive to Zn deficiency when exposed to long sunny days and drought conditions, probably due to enhanced photooxidative damage in leaves with low Zn concentrations [46]. The differential tolerance to salt stress among wheat genotypes was highly related to their tolerance to Zn starvation; particularly, greater tolerance to high salinity-induced Zn deficiency was probably associated with significant differences in antioxidant defence capacity among them (significantly greater root activity of catalase and superoxide dismutase) [27, 45]. According to Cakmak *et al.* (1997a), the sensitivity of cereal species and cultivars to show the Zn deficiency symptoms was closely related with the Cu/Zn SOD activities, and partially with that of total SOD in the leaf tissues [44].

III. **Enhanced mobilization and translocation of Zn among plant tissues.** Roots often contain much more Zn than do tops, particularly if the plants are grown in Zn-rich soils. However, even under Zn deficiency conditions several plant species are able to mobilize limited- but crucial for plant growth- Zn quantities from older to younger leaves [21]. Enhanced mobilization and translocation from root system (which represents a valuable storage place containing an important 'pool' of Zn) to top parts is another greatly important adaptive mechanism of plants that substantially help them to overcome this nutritional stress.

IV.   **The formation of mycorrhiza.** It is well known that under different stress conditions plants are able to form mycorrhiza in order to survive. This symbiosis between fungus and plant is mutually beneficial for both parts, since plant may supply fungus with carbohydrates produced by photosynthesis and fungus enhances nutrient uptake, which is crucial for plant survival under nutrient stress conditions. Especially, the formation of arbuscular mycorrhizas (AM) can increase Zn plant uptake under low soil concentrations [60]. Arbuscular mycorrhiza fungus (AMF) colonization in some studies was found to be influenced by Zn level [61], while in some others not [62]. This is not unexpected, since there are limited studies on this topic, a diverse range of plants, soils, AMF and Zn treatments directly addressing this issue [60].

Zn Fertilization

It has been calculated that Zn uptake by crops is usually less than 0.5 kg/ha/season. In practice, it is easy to correct Zn deficiency, either by spraying, or by soil application with Zn fertilizers [13]. The mostly used salt to alleviate Zn deficiency is $ZnSO_4.7H_2O$ (it contains 22.7% Zn). It is highly soluble and it may be applied either as soil or spray fertilizer. On acid sandy soils it may be preferable to spray the crop or use a less readily available Zn source because $ZnSO_4$ is very easily leached. The same applies to alkaline soils, which fix Zn very strongly [13]. In *Citrus* plants for example, it may be applied as spray fertilizer, a little before the beginning of new vegetation, or in early summer. In apple trees (*Malus domestica*), one spraying with 2.5-5% solution at the end of the winter period (a little before bud swelling) is referred to be enough. In *Zea mays* plants, a quantity of 3-4 kg of that salt is usually applied during sowing [16]. In rice plants, more distinct increases in grain Zn by foliar Zn applications were achieved when Zn was applied after flowering [19]. Zn-EDTA is an organic chelate containing 14% Zn that can be applied either as soil fertilizer, or spray in leaves; after soil application, irrigation is absolutely necessary. A quantity of 0.5 to 1 kg per tree (about 500 g. per tree for Avocado trees and 0.9-1 kg per tree for *Prunus amygdalus*) is referred as the most appropriate for the correction of Zn deficiency and stimulation of plant growth and fruit setting. In apple trees (*Malus domestica*) Zn is usually applied in solution 0.2% as spray in leaves during their early appearance [16]. According to Ozkutlu *et al.* (2006), Zn humates and $ZnSO_4$ were found to be similarly effective in increasing dry matter production in wheat; but Zn humates increased soybean dry matter more than $ZnSO_4$. So, their results indicated that soybean and wheat plants can efficiently utilize Zn chelated to humic acids in calcareous soils [49].

Apart from $ZnSO_4$, which has been traditionally used as the reliable source of Zn fertilizer, other sources of Zn are also available. Some are derived from industrial by-products, varying from flue dust reacted with sulphuric acid to organic compounds derived from the paper industry. The degree of Zn mobility in Zn sources derived from these various by-products is related to the manufacturing process, the source of complexing or chelating agents (organic sources), and the original product used as Zn source. More specifically, apart from ZnEDTA and $ZnSO_4$, which are very highly water soluble, ZnLigno, which is formed by reacting $ZnSO_4$ with lignin wastes produced by the paper industry, is a very good source of Zn (containing 10% total Zn) and highly water-soluble (91% of the total Zn is water soluble) [7].

It was found that foliar application alone, or in combination with soil application, significantly increased grain Zn concentrations from 27 mg/kg d.w. to 49 mg/kg d.w. in 23 sites from different countries. Foliar Zn fertilizer approach can be locally adopted for increasing dietary Zn intake and fighting human Zn deficiency in rural areas of developing countries [37]. Phattarakul *et al.* (2012) found that Zn fertilization had little effect on rice grain yield, in rice plants grown in different Asian countries (as an average Zn application increased grain yield by 5%); grain Zn concentrations were, however, more effectively increased by Zn fertilization, especially with foliar applications (25% by foliar application, 32% by foliar+soil application, and only 2.4% by soil Zn application) [19]. According to Manzeke *et al.* (2012), a combined use of organic and inorganic fertilizer yielded >2.1 tn/ha maize grain, against <0.8 tn/ha in the non fertilized (control). Finally, maize grain Zn concentrations increased by 46-64% over the control [18].

REFERENCES

[1]   Kochian LV. Molecular physiology of mineral nutrients acquisition, transports and utilization. In: Buchanan BB, Gruissem W, Jones RL, Eds. Biochemistry and molecular biology of plants. American Society of Plant Biologists, Rockville, 2000; pp 1204-1249.

[2]   Cakmak I. Enrichment of fertilizers with zinc: an excellent investment for humanity and crop production in India. J Trace Elem Med Biol 2009; 23: 281-289.

[3]   Parker DR, Aguilera JJ, Thomason DN. Zinc-phosphorus interactions in two cultivars of tomato (*Lycopersicon esculentum* L.) grown in chelator-buffered nutrient solutions. Plant Soil 1992; 143: 163-177.

[4]   Rashid A, Ryan J. Micronutrient constraints to crop production in soils with Mediterranean type characteristics: A review. J Plant Nutr 2004; 27: 959-975.

[5]   Ozkutlu F, Torum B, Cakmak I. Effect of zinc humate on growth of soybean and wheat in zinc-deficient calcareous soil. Com Soil Sci Plant Anal 2006; 37: 2769-2778.

[6]   Cakmak I. Enrichment of cereal grains with zinc: agronomic or genetic biofortification? Plant Soil 2008; 302: 1-17.

[7]   Gangloff WJ, Westfall DG, Peterson GA, Mordvedt JJ. Mobility of organic and inorganic zinc fertilizers in soils. Com Soil Sci Plant Anal 2006; 37: 199-209.

[8]   Cakmak I, Gulut KY, Marschner H, Graham RD. Effect of zinc and iron deficiency on phytosiderophore release in wheat genotypes differing in zinc deficiency. J Plant Nutr 1994; 17: 1-17.

[9]   Peng GX, Fusuo Z, Hoffland E. Malate exudation by six aerobic rice genotypes varying in zinc uptake efficiency. J Environ Qual 2009; 38: 2315-2321.

[10] Kothari SK, Marschner H, Romheld V. Direct and indirect effects of VA mycorrhizal fungi and rhizosphere microorganisms on acquisition of mineral nutrients by maize (*Zea mays* L.) in a calcareous soil. New Phytol 1990; 116: 637-645.

[11] Cakmak I, Hoffland E. Zinc for the improvement of crop production and human health. Plant Soil 2012; 361: 1-2.

[12] Cakmak I, Kalaysi M, Ekiz H, Brawn HJ, Yilmaz A. Zinc deficiency as an actual problem in plant and human nutrition in Turkey: a NATO-science for stability project. Field Crops Res 1999; 60: 175-188.

[13] Mengel K, Kirkby E. Zinc. In: Mengel K, Kirkby E, Kosegarten H, Appel T, Eds. Principles of Plant Nutrition, 5th Edition. Kluwer Academic Publishers, Dordrecht, The Netherlands 2001; pp 585-597.

[14] Lindsay WL, Norvell WA. Development of a DTPA soil test for zinc, iron, manganese and copper. Soil Sci Soc Am J 1978; 42: 6421-6428.

[15] Takkar PN, Chhibba IM, Mehta SK. Twenty years of coordinated research on micronutrients in soils and plants. Bulletin of Indian Institute of Soil Science 1989; 1: 76.

[16] Therios I. Mineral Nutrition and Fertilizers. Dedousis Publications, Thessaloniki, Greece 1996; pp. 174-177. (In Greek).

[17] Peck AW, McDonald GK. Adequate zinc nutrition alleviates the adverse effects of heat stress in bread wheat. Plant Soil 2010; 337: 355-374.

[18]   Mankeze GM, Mapfumo P, Mtambanengwe F, Chikowo R, Tendayi T, Cakmak I. Soil fertility management effects on maize productivity and grain zinc content in smallholder farming systems of Zimbabwe. Plant Soil 2012; 361: 57-69.

[19]   Phattarakul N, Rerkasem B, Li LJ, *et al*. Biofortification of rice grain with zinc through zinc fertilization in different countries. Plant Soil 2012; 361: 131-141.

[20]   Rashid A, Rafique E, Bughio N, Yasin N. Micronutrient deficiencies in rainfed calcareous soils of Pakistan. IV. Zinc nutrition of sorghum. Com Soil Sci Plant Anal 1997; 28: 455-467.

[21]   Kabata-Pendias A, Pendias H. Trace Elements in Soils and Plants. 3rd ed. CRC Press, USA, 2001.

[22]   Catlett KM, Heilb DM, Lindsayc DM, Ebingerd MH. Soil chemical properties controlling Zn^{2+} activity in 18 Colorado soils. Soil Sci Soc Am J 2002; 66: 1182-1189.

[23]   Marschner H. Zinc uptake from soils. In: Robson AD, Ed. Zinc in soils and plants. Kluwer Academic publishers: Dordrecht, The Netherlands 1993; pp. 59-77.

[24]   International Zinc Nutrition Consultative Group. Improving health of people in need by enhancing zinc nutrition. Internet: http://www.izincg.org/

[25]   Alifragis D. Soil: Genesis, Properties and Classification. Volume I. Aivazi Publications, Thessaloniki, Greece 2008; pp. 487-492. (In Greek).

[26]   Mortvedt JJ, Gilkes RJ. Zinc fertilizers. In: Robson AD, Ed. Zinc in soils and plants. Kluwer Academic Publishers, Dordrecht, The Netherlands, 1993.

[27]   Khoshgoftarmanesh AH, Shariatmadari H, Karimian N. Responses of wheat genotypes to zinc fertilization under saline soil conditions. J Plant Nutr 2006; 29: 1543-1556.

[28]   Rego TJ, Sahrawat KL, Wani SP, Pardhasaradhi G. Widespread deficiencies of sulphur, boron and zinc in Indian semi-arid tropical soils: On farm crop responses. J Plant Nutr 2007; 30: 1569-1583.

[29]   Abd-Elfattah A, Wada K. Adsorption of lead, copper, zinc, cobalt and cadmium by soils that differ in cation-exchange materials. J Soil Sci 1981; 32: 271-283.

[30]   Cakmak I, Hoffland E. Zinc for the improvement of crop production and human health. Plant Soil 2012; 361: 1-2.

[31]   Kutman UB, Kutman BY, Ceylan Y, Ova EA, Cakmak I. Contributions of root uptake and remobilization to grain zinc accumulation in wheat, depending on post anthesis, zinc availability and nitrogen nutrition. Plant Soil 2012; 361: 177-187.

[32]   Marschner H. Mineral Nutrition of Higher Plants, 2nd edition. Academic Press, London 1995.

[33]   Graham RD, Welch RM, Grunes DL, Cary EE, Norvell WA. Effects of zinc deficiency on the accumulation of boron and other mineral nutrients in barley. Soil Sci Soc Am J 1987; 51: 652-657.

[34]   Daneshbakhsh B, Khoshgoftarmanesh AH, Sariatmadari H, Cakmak I. Phytosiderophore release by wheat genotypes differing in zinc deficiency tolerance grown with Zn-free nutrient solution as affected by salinity. J Plant Physiol 2013; 170: 41-46.

[35]   Citernesi AS, Vitagliano C, Giovannetti M. Plant growth and root system morphology of *Olea europaea* L. rooted cuttings as influenced by arbuscular mycorrhizas. J Hortic Sci Biotech 1998; 73: 647-654.

[36]   Chatzistathis T, Orfanoudakis M, Alifragis D, Therios I. Colonization of Greek olive cultivars' root system by arbuscular mycorrhiza fungus: root morphology, growth, and mineral nutrition of olive plants. Sci Agric 2013; 70: 185-194.

[37]   Zou CQ, Zhang YQ, Rashid A, *et al*. Biofortification of wheat with zinc through zinc fertilization in seven countries. Plant Soil 2012; 361: 119-130.

[38]   Suzuki M, Takahashi M, Tsukamoto T, *et al*. Biosynthesis and secretion of mugineic acid family phytosiderophores in zinc-deficient barley. Plant J 2006; 48: 85-97.

[39]   Suzuki M, Tsukamoto T, Innoue H, *et al*. Deoxymugineic acid increases Zn translocation in Zn-deficient rice plants. Plant Mol Biol 2008; 66: 609-617.

[40]   Ishimaru Y, Bashir K, Nishizawa NK. Zn uptake and translocation in rice plants. Rice 2011; 4: 21-27.

[41]   Palmgren MG, Clemens S, Williams LE, Kramer U, Borg S, Schjorring JK, Sanders D. Zinc biofortification of cereals: problems and solutions. Trends Plant Sci 2008; 13: 464-473.

[42]   Bashir K, Ishimaru Y, Nishizawa NK. Molecular mechanisms of zinc uptake and translocation in rice. Plant Soil 2012; DOI 10.1007/s11104-012-1240-5

[43]   Assuncao AGL, Schat H, Aarts MGM. Regulation of the adaptation to zinc deficiency in plants. Plant Signal Behav 2010; 5: 1553-1555.

[44]   Cakmak I, Ozturk L, Eker S, Torun B, Kalfa HI, Yilmaz A. Concentration of zinc and activity of copper/zinc superoxide dismutase in leaves of rye and wheat cultivars differing in sensitivity to zinc deficiency. J Plant Physiol 1997a; 151: 91-95.

[45]   Daneshbakhsh B, Khoshgoftarmanesh AH, Shariatmadari H, Cakmak I. Effect of zinc nutrition on salinity-induced oxidative damages in wheat genotypes differing in zinc deficiency tolerance. Acta Physiol Plant 2012; DOI 10.1007/s11738-012-1131-7

[46]   Cakmak I. Possible roles of Zn in protecting plant cells from damage by reactive oxygen species. New Phytol 2000; 146: 185-205.

[47]   Chanakan PT, Rerkasem B, Yazici A, Cakmak I. Zinc priming promotes seed germination and seedling vigor of rice. J Plant Nutr Soil Sci 2012; 175: 482-488.

[48]   Zinc deficiency in a strawberry plant. By J. Brennan. Internet: http://www.ehow.com/info_8776902_zinc-deficiency-strawberry-plant.html

[49]   Ozkutlu F, Torum B, Cakmak I. Effect of zinc humate on growth of soybean and wheat in zinc-deficient calcareous soil. Com Soil Sci Plant Anal 2006; 37: 2769-2778.

[50]   Cakmak I, Ekiz H, Yilmaz A, Torun B, Koleli N, Gultekin I, Alkan A, Eker S. Differential response of rye, triticale, bread and durum wheat to zinc deficiency in calcareous soils. Plant Soil 1997b; 188: 1-10.

[51]   Trace Elements. Oligo Zinc-Nitrate 42% Liquid. Internet: http://www.vaniperen.com/Products/Trace-elements/Non-chelated/Oligo-Zinc-Nitrate-42--Liquid.aspx

[52]   Epstein E, Bloom AJ. Mineral Nutrition of Plants: Principles and Perspectives. 2nd ed. 2005.

[53]   Nutrient disorder management in cotton. Zinc deficiency. Internet: http://jnkvv.nic.in/IPM%20Project/nutrient-cotton.html#ZincDeficiency

[54]   Zinc deficiency symptom in lettuce. Internet: http://www.rqsulfates.com/newsr_3_6_0_762.html

[55]   Rengel Z, Graham RD. Uptake of zinc from chelate-buffered nutrient solutions by wheat genotypes differening in zinc efficiency. J Exp Bot 1996; 47: 217-226.

[56]   Graham R, Senadhira D, Beebe S, Iglesias C, Monasterio I. Breeding for micronutrient density in edible portions of staple food crops: conventional approaches. Field Crops Res 1999; 60: 57-80.

[57]   Wu CY, Lu LL, Yang XE, *et al.* Uptake, translocation and remobilization of zinc absorbed at different growth stages by rice genotypes of different Zn densities. J Agric Food Chem 2010; 58: 6767-6773.

[58]   Peng GX, Fusuo Z, Hoffland E. Malate exudation by six aerobic rice genotypes varying in zinc uptake efficiency. J Environ Qual 2009; 38: 2315-2321.

[59]   Cakmak I, Sari N, Marschner H, *et al.* Phytosiderophore release in bread and durum wheat genotypes differing in zinc efficiency. Plant Soil 1996; 180: 183-189.

[60]   Cavagnaro TR, Dickson S, Smith FA. Arbuscular mycorrhizas modify plant responses to soil zinc addition. Plant Soil 2010; 329: 307-313.

[61]   Ortas I, Ortakei D, Kaya Z, Cinar A, Onelge N. Mycorrhizal dependency of sour orange in relation to phosphorus and zinc nutrition. J Plant Nutr 2002; 26: 1263-1279.

[62]   Chen BD, Li XL, Tao HQ, Christie P, Wong MH. The role of arbuscular mycorrhiza in zinc uptake by red clover growing in a calcareous soil spiked with various quantities of zinc. Chemosphere 2003; 50: 839-846.

CHAPTER 5

Manganese Deficiency

Abstract: Despite the fact that Mn is one of the most abundant trace elements in the lithosphere, its' concentration in soils greatly varies among soil types, depending on parent material and soil conditions. Apart from the parent material, other soil factors influencing Mn availability are pH, organic matter, soil moisture, $CaCO_3$ content, redox potential, soil microorganisms, soil texture, the availability of other nutrients, the kind of N fertilizers (NO_3^- or NH_4^+) used by the farmers *etc*.

Manganese uptake by plants is metabolically controlled, *i.e.*, it is absorbed as Mn^{2+}, with energy consumption. In some cases Mn may be also taken up as chelate. Manganese is highly important for plant metabolism and growth, since it is contained in three enzyme complexes (those of photosystem II, MnSOD and acid phosphatases), while it is also an activator for a large number of enzymes (about 35). Manganese plays a very important role in photosystem II and photosynthesis, while it is also associated with the structure of chloroplasts, with N metabolism, with CO_2 assimilation in C4 plants, with the cycle of tricarboxylic acids *etc*. Critical Mn deficiency concentrations for plant growth vary from 15 to 20 mg/kg d.w. (depending on plant species). The most characteristic symptoms of Mn deficiency are the appearance of small yellow spots on the leaves and the interveinal chlorosis. In cases of Mn deficiency symptoms, soil or foliar application of Mn (either as $MnSO_4$ or MnEDTA) should be preferred. Particularly, foliar application of Mn is advantageous in calcareous/alkaline soils, where Mn is quickly immobilized after soil application.

Under conditions of Mn deficiency plants adopt tolerance mechanisms in order to survive. Some of the most important mechanisms include the acidification of their rhizosphere, the greater ability for remobilization and redistribution of Mn from the more to less tolerant tissues (such as the young leaves), the formation of arbuscular mycorrhiza to root system environment, the adjustment of root morphology in order to take up more Mn^{2+} *etc*. From an agronomical point of view, under conditions of limited Mn availability particular emphasis should be given to the choice and cultivation of genotypes with increased ability of Mn uptake and internal use efficiency (*e.g.*, enhanced transport from the root system to shoot).

All these topics concerning Mn content and availability in soils, Mn uptake by plants, the role of Mn in biochemical and physiological reactions, the symptoms of Mn deficiency, as well as the strategies adopted by plants in order to survive under Mn starvation and the adaptation mechanisms of the Mn-use efficient genotypes are fully discussed in this chapter.

Keywords: Acid phosphatases, Mn availability, Mn deficiency, Mn fertilizers, Mn tolerance, Mn uptake, Mn utilization efficiency, MnEDTA, MnSOD, $MnSO_4$, photosynthesis, photosystem II.

Theocharis Chatzistathis

INTRODUCTION

Manganese is one of the most abundant trace elements in lithosphere and its common range in rocks varies from 350 to 2000 p.p.m. [1]. According to other authors, Mn concentration of the upper part of the earth's crust is on average 650 p.p.m. [2]. Lindsay (1979) refers that the mean Mn concentration of the lithosphere is about 900 p.p.m. [3]. There are many Mn containing minerals (different oxides-hydroxides, carbonate, siliceous, sulphureous salts); however, the most important are those of oxides-hydroxides, which have highly adsorptive capacity for heavy metals due to their negative charge. Some of the most important Mn oxides-hydroxides, which are formed by the 'released' Mn, contained in differents rocks by weathering, are basically pyrolusite, manganite (secondary minerals), as well as birnesite, hollandite *etc.* [4, 5]. Although Mn in soils could be present under three different forms (Mn^{2+}, Mn^{3+} and Mn^{4+}), only the first one (divalent) is directly available for plants (it is present in soil solution and can by absorbed) and contributes to mineral nutrition. Generally, there is a dynamic equilibrium in soils between Mn^{4+} and Mn^{2+} (this equilibrium is influenced by pH, soil moisture, $CaCO_3$ content, organic matter, redox potential, soil microorganisms, soil texture, the availability of other nutrients *etc.*). Most of these soil factors, together with other ones, directly influenced by plants and human activities, may influence not only Mn solubility and availability, but also its uptake. Apart from the Mn^{2+} form, the complexes of Mn with organic ligands may contribute to Mn solubility and availability, especially in the alkaline pH range [1].

Manganese uptake is metabolically controlled; after Mn^{2+} being absorbed by plants with energy consumption, it is transferred with the transpiration stream towards shoots and leaves. In contrast to that, it was found that Mn is relatively immobile inside plants, since it is scarcely translocated in phloem [5]. Despite the fact that Mn is required in very low concentrations in plants, it is highly important for plant metabolism and growth, since it is contained in three enzyme complexes (those of photosystem II, MnSOD and acid phosphatases), while it is also an activator for a large number of enzymes (about 35) [6]. The most important role of Mn is its' participation in the O_2-evolving system of photosynthesis and in the transfer of electrons in the photosynthetic electron transport chain. Due to the great importance that Mn plays in photosystem II, photosynthetic rates may be disturbed under Mn deficiency [7-9]. Manganese is also associated with the structure of chloroplasts, the assimilation of N, with CO_2 assimilation in C4 plants, with the cycle of tricarboxylic acids *etc.* [1, 5]. Under conditions of Mn starvation, Mn deficiency symptoms may appear in plants; some of the most typical ones are those of interveinal chlorosis, yellow or necrotic spots on leaves, structural impairment and disorganization of chloroplasts *etc.* When leaf Mn concentration in crops is lower than 20 mg/kg d.w. plants suffer from Mn deficiency and soil or foliar Mn applications are required in order to correct this nutritional disorder. Generally, $MnSO_4$ is superior to Mn chelates for soil application and most effective for foliar application, while from the organic complexes Mn-EDTA appears to be the most promising organic ligand, depending also on plant species and genotype characteristics [5].

The purposes of this chapter are: i) to present all the most important data concerning Mn in soils, as well as to analyze the way by which different soil, plant and other factors influence Mn availability and uptake, ii) to describe the roles of Mn in plant physiology, biochemistry and metabolism, iii) to provide all the important information concerning Mn deficiency symptoms, critical concentrations for crops, genotypic tolerance and Mn fertilizers (*via* soil or foliar application), iv) to highlight the mechanisms adopted by plants in order to survive under conditions of Mn starvation, and finally, v) to provide recommendations for Mn fertilization of crops.

Mn Content of Different Rocks and Mn Containing Minerals

The Mn concentration of different rocks varies a lot; generally, the greatest Mn contents have been found in basic-eruptive rocks (basalt, gabbro), varying between 1000 and 2000 p.p.m. Manganese contents vary also widely in acid eruptive (granite, rhyolite), metamorphic (schists), as well as in certain sedimentary rocks, varying among 200 and 1200 p.p.m. Average contents have been found in limestones (400-600 p.p.m.), while in sands Mn contents are relatively low (20-500 p.p.m.) [10]. According to Gilkes and McKenzie (1988), the representative Mn concentration in igneous rocks (basalt, andesite, rhyolithe, phonolite, peridotite, gabbro, anorthosite, diorite, granite *etc.*) varies from 200 p.p.m. (anorthosite) to 1500 p.p.m. (in phonolite) [2]. Basalt representative concentration is around 1300 p.p.m., while those of gabbro and andesite are around 1200 p.p.m. In metamorphic rocks Mn representative concentrations were found to vary from 600 to 1500 p.p.m. (Gneiss concentration is around 600 p.p.m. and amphibolites on average 1500 p.p.m.). Finally, in sedimentary rocks, Mn content varies a lot; sandstone concentrations were found to be approximately, 170 p.p.m., while deep sea clay concentrations were on average 6000 p.p.m. [2]. According to Kabata and Pendias (2001), Mn common range in rocks varies from 350 to 2000 p.p.m. [1], while Lindsay (1979), refers that the mean Mn concentration of the lithosphere is about 900 p.p.m. [3]. The great variation of Mn content between the different rocks is responsible for the wide variation of Mn content of soils: from traces (podzols of Poland) to 10000 p.p.m. (in the unleached alkali soils of Chad) [10]. Table **1** shows the mean Mn concentrations of the most important rock types.

From the different minerals containing Mn (different oxides-hydroxides, carbonate, siliceous, sulphureous salts), the most important are those of oxides-hydroxides, which have highly adsorptive capacity for heavy metals due to their negative charge. Taylor and McKenzie (1966) refer that Co in many cases is adsorbed by Mn oxides-hydroxides; this is of great importance for agriculture, since Co availability for plants depends on the abundance of Mn oxides-hydroxides in soils [11]. Only a moderate abundance of Mn oxides-hydroxides could be responsible for limited Co availability for plants and deficiency; the Co adsorbed by Mn oxides becomes more strongly bound with increasing time and it is oxidized and replaces Mn in the crystal structure [2].

Some of the most important Mn oxides-hydroxides are birnesite, hollandite, pyrolousite, manganite *etc.* [4]. The negatively charged surfaces of Mn oxides and hydroxides are responsible not only for the adsorption of Co, but also for the high degree of association between them and Ni, Cu, Zn, Pb, Ba, Mo *etc.* [1]. According to Gilkes and McKenzie (1988), the specific adsorption of heavy metal cations by Mn oxides occurs in the order Pb>Cu>Mn>Co>Zn>Ni [2]. McKenzie (1977) refers that Mn is likely to occur in soils as oxides and hydroxides in the form of coatings on other soil particles and as nodules of different diameters; these nodules often exhibit a concrete layering that is suggestive of seasonal growth [4]. Manganese oxides in soils are mostly amorphous, but crystalline varieties have been also identified in several soils [1].

Table 1.   **Manganese concentrations in major rock types (modified from [2]).**

Rock Type	Representative Mn Concentration (p.p.m.)
a) Igneous Rocks	
Basalt	1300
Andesite	1200
Rhyolithe	600
Phonolite	1500
Peridotite	1200
Gabbro	1200
Anorthosite	200
Granite	400
b) Metamorhic Rocks	
Gneiss	600
Granulite	800
Amphibolites	1500
c) Sedimentary Rocks	
Sandstone	170
Limestone	550
Deep sea clays	6000

Mn Oxidation and Reduction in Soils

The basic Mn forms in soils are the following: a) the soluble Mn (that of soil solution, *i.e.*, Mn^{2+}), b) the exchangeable Mn (that which is adsorbed in soil colloids and can be substituted by other divalent cations and 'liberated' in soil solution), c) the organically bound soluble Mn (that which can be absorbed by plants as a complex with organic matter), d) the easily reducible Mn, e) the inert and insoluble Mn (that of the lattice of minerals, as well as that contained in Mn oxides-hydroxides), f) the insoluble complexes of Mn with organic matter [12]. From the above mentioned Mn forms, only the soluble (Mn^{2+}) and the exchangeable Mn are the ones that can be directly taken up by plants. The basic

Mn sources in soils are the different minerals, by which through the processes of weathering, hydrolysis and dilution, Mn^{2+} is 'liberated' in soil solution (a form that can be absorbed by plants). There is an equilibrium between soluble Mn^{2+} and insoluble (inert) Mn^{4+} in soils, according to the following reaction:

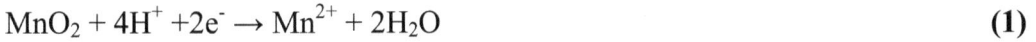

$$MnO_2 + 4H^+ + 2e^- \rightarrow Mn^{2+} + 2H_2O \qquad (1)$$

Many factors influence the transformation of Mn^{4+} to soluble Mn^{2+}, a form that can be absorbed by plants. Conditions of good soil aeration and neutral or alkaline pH are the most important factors favoring the prevalence of insoluble Mn (Mn^{4+}). In contrast to that, when soils are humid/or waterlogged and soil oxygen is poor, Mn is reduced from Mn^{4+} to Mn^{2+}, according to the equation (1). This reaction may be also observed in wet soils, which have not, however, reached their saturation point of soil moisture. According to Graven *et al.* (1965), 72 hours after reaching the saturation point in soil moisture, Mn concentration of *Medicago sativa* L. plants has been increased from 426 p.p.m. to more than 6000 p.p.m. [13]. Finally, Mandal (1961) found that Mn oxides in soils, cultivated with *Oryza sativa* plants, have been dramatically decreased after the saturation of these soils with water [14].

Factors which Influence the Concentration of Available Mn and its Uptake by Plants

There are many soil factors influencing the concentration of soluble Mn (that of Mn^{2+}) in soils, thus its' uptake by plants. The most important of these factors are pH, organic matter and $CaCO_3$ content, soil moisture, redox potential, soil texture, soil microorganisms, the phosphoric ion content, the content of soils in Fe and Al oxides-hydroxides, the interaction with other nutrients *etc*. Apart from the soil factors determining plant available Mn, some plant and other factors, influenced by human activities, determine also the concentration of soluble Mn in soil. From the plant factors, they should be distinguished root exudations and the formation of mycorrhiza, while from the factors influenced by human activities the most important are the kind of N fertilization (NO_3^- or NH_4^+) used and the choice of the suitable plant species/genotypes/cultivars for cultivation. All the above-mentioned factors are described and fully analyzed below.

Soil Factors

I.   **pH.** Generally, Mn solubility is increased when soil pH is decreased. Manganese uptake is closely related to soil pH, more than that of other micronutrients [15]. Generally, the highest Mn availability occurs in pH 5-5.5 [16].

II. **Soil moisture and redox potential.** In wet or waterlogged soils, Mn^{4+} is converted into Mn^{2+} (a form that can be absorbed by plants) due to the lack of oxygen. In dry soils Mn is prevalent in insoluble, non-available for plants form (Mn^{4+}). Soil moisture (thus Mn reduction) is closely associated with redox potential in soils; according to Grass *et al.* (1973), the reduction of Mn takes place when the redox potential is lower than 300 mV [17]. Patrick and Turner (1968) found that the transformation of the easily reducible into the exchangeable Mn begins taking place in redox potential values around 400 mV [18]. There is an interaction between pH and redox potential in Mn solubilization. According to Gotoh and Patrick (1972), in a soil with pH 5, saturated with water, almost all the quantity of the reducible Mn converted into soluble and exchangeable in redox potential values around 500 mV. In contrast to that, when pH values were between 6 and 8, the greatest part of the reducible Mn was converted into soluble and exchangeable Mn in much lower redox potentials (200-300 mV) [19].

III. **The organic matter content.** The different metals have the ability to form with organic matter organometallic complexes. Usually, these complexes are insoluble, thus their formation decreases metal (also Mn) availability for plants. However, there are cases when organic Mn complexes constitute about 80-95% of the total soluble Mn forms; in addition, their stability is usually increased with the increase of soil pH. So, the importance of organic Mn complexes for mineral nutrition of plants is great under alkaline soil conditions [16]. Consequently, as in the case of Fe, alkaline conditions do not necessarily provoke Mn deficiency. Wei *et al.* (2006) found that the decomposition of organic matter provided protons to the soil solution and resulted in the dissolution and reduction of Mn, thus increasing its availability in some soils after 18 years of continuous cropping and fertilization [20].

IV. **CaCO₃ content.** When $CaCO_3$ content increases, Mn solubility and availability decrease. Thus, under high $CaCO_3$ content plants may suffer from Mn starvation due to decreased Mn uptake. Apart from the direct influence of $CaCO_3$ on the decrease of Mn solubility, another reason for the reduction of Mn uptake under high $CaCO_3$ conditions can be the competition between Ca and Mn for uptake by plants [21].

V. **Soil texture.** Soil texture influences Mn availability through leaching, *i.e.*, in sandy soils Mn is leached easier, than in clayey ones [22].

VI. **Soil microorganisms.** Manganese oxidation may take place through biological processes (basically through aerobic bacteria). These bacteria are not very sensitive to soil acidity, but they are very active in a pH range from 6 to 7.5 [12]. According to Douka (1973), bacteria of the genus *Pseudomonas* and *Citrobacter*, found in western Peloponnesus, Greece, are capable of oxidizing, through enzymatic or non-enzymatic processes, Mn^{2+} into Mn oxides (Mn^{4+}) [23]. Other kind of microorganisms which may oxidize Mn^{2+} into Mn oxides are the bacteria of genus *Arthrobacter sp.*, which are particularly active in a range of pH varying between 5.7 and 7.5 [15]. The most usual genus of bacteria, which are responsible for Mn oxidation, are those belonging to *Arthrobacter, Bacillus, Pseudomonas* and *Hyphomicrobium*. Other, less frequent, occuring genuses are those of *Corynebacterium, Proteus, Flavobacterium, Enterobacter* and *Citrobacter*. From the fungi responsible for Mn oxidation we should distinguish those of the genus *Streptomyces,* as well as that of *Gaeumannomyces graminis* [24].

VII. **Soil temperature.** Manganese uptake is influenced by soil temperature and its' deficiency is more likely to take place under low soil temperatures. It seems that Mn uptake is influenced by soil temperature, more than that of Fe [15].

VIII. **The phosphoric ion content of soils.** High phosphoric ion content in soils may lead to the formation of insoluble Mn-phosphate substances [22]. In that case, Mn availability is significantly decreased and plants may suffer from Mn deficiency.

IX. **The concentration of other nutrients, which may act synergistically or competitively to Mn absorption.** Many competitive or synergistic effects of several nutrients (such as Fe, Ca and Mg) on Mn uptake by plants have been referred. The most important is the interaction between Mn and Fe. Clarkson (1988) supports that Mn and Fe act competitively for a place of absorption in root surface [25]. According to El-Jaoual and Cox (1998), the nature of interaction between Mn and Fe differs among plant species and while in most of them is competitive, for some others is neutral, or even synergistic [21]. We found that not only plant species, but also genotype of the same species (olive cultivars) influenced the nature of interaction between Fe and Mn [26]. According to Ghasemi-Fasaei *et al.* (2002), high Fe supply in soil may reduce Mn uptake from plants and its transport from root to shoot [27]. Increased levels of Ca in soil, or nutrient, solution have a negative impact on Mn uptake by plants. Very often, of great importance is the ratio Ca/Mn. In *Pistacia vera*, this ratio for optimum mineral nutrition should be greater than 80 [21]. Increased Mg concentrations in soils act competitively to Mn uptake by plants [28, 29]. There are contradictory results concerning the influence of some metals, such as Cu and Zn, on Mn uptake by plants [21, 30, 31]; some researchers found that these effects were antagonistic, whilst some others supported that these interactions were synergistic. It seems that plant species, as well as genotypes of the same species, determine these effects. Finally, it was found that reduction in the leaf Mn concentration by B application may be due to the dilution effect or to the antagonistic relationship between B and Mn [32].

Non-Soil Factors (Plant Factors and Other Ones, Influenced by Agricultural Practices)

I. **The use of heavy agricultural machine equipment.** The structure of wet/waterlogged clayey soils may be considerably deteriorated when heavy agricultural machine equipment is used for crop production, thus bad aeration of deeper soil horizons may occur. In that case, reduction of Mn^{4+} to Mn^{2+} may be a reality.

II. **Soil disinfection.** Chemical soil disinfection is a very important agronomical practice, adopted by many farmers in greenhouses. After soil chemical disinfection, Mn reduction takes place in neutral pH soils [33, 34]. The enhancement of Mn reduction after soil disinfection may be ascribed to one of the two following reasons (or to both of them): to the release of the organically bound Mn, and/or to the death of the microorganisms, which are responsible for the oxidation of Mn^{2+} to Mn^{4+} [35].

III. **Root exudations.** According to Smith and Paterson (1990), who studied the fluctuation of Mn concentration in soil solution, in a soil cultivated with *Hordeum vulgare*, a maximum of Mn concentration was found during the early summer period, together with a slight increase of pH. A possible explanation for this phenomenon was the production of low molecular weight organic substances, produced by the roots during the period of the maximum growth rate of plants. These organic substances acted as a factor of creation of complexation between them and Mn, increasing its availability to plants [12]. Bromfield (1958) found that the root exudations of 20 week-old *Avena sativa* plants were able to solubilize Mn oxides, and their solubilization ability was inversely correlated with soil pH (the quantity of Mn solubilized in pH values greater than 6.5 was very small) [36].

IV. **Mycorrhiza formation.** Generally, Mn is one of the nutrients that its uptake is mostly influenced by mycorrhiza. Generally, mycorrhiza formation significantly contributes to the uptake of Mn^{2+}, while it is also referred that it enhances the uptake of Mn-organic complexes [16]. Nevertheless, in inoculated with *Glomus mosseae* maize plants the ability of root exudations for Mn reduction was two times lower, compared to that of the control (non inoculated with *Glomus mosseae*) ones [37]. It has been also found that the proportion of microorganisms which are responsible for

Mn uptake was 20-30 times greater in non-inoculated with *Glomus mosseae* maize plants, than in inoculated plants [38].

V.   **The kind of N fertilization.** Plants receiving NH_4 fertilization acidify their rhizosphere and increase the availability of Mn (and therefore, its uptake by plants) [39], in contrast to those receiving NO_3 fertilization [17].

VI.   **The use of irrigation water with high electrical conductivity and high bicarbonate ion content.** According to Somasundaram *et al.* (2011), the use of irrigation water with high electrical conductivity and high bicarbonate ion content resulted in a temporary raise of soil pH, which leaded to reduced Mn availability to *Citrus* trees [40].

VII.   **The ability of plant species and genotypes to absorb Mn.** Among the different plant species, great differences exist concerning Mn uptake. According to Graham (1988), from a great variety of plant species being susceptible to Mn deficiency, the most important are those of *Avena sativa*, *Citrus* plants, *Malus domestica*, *Prunus* sp. and several species of Brassicaceae family [41]. Apart from the differential Mn uptake ability existing between plant species, great genotypic differences may also exist among different cultivars of the same species [42-47]. These great genotypic differences may be probably ascribed to the differential root growth response under Mn starvation, or to the differential ability for root exudations (organic exudates which may solubilise Mn and enhance its' uptake in the form of organic complexes).

Determination of Plant Available Mn in Soils

Plant available Mn in soils is controlled not just by soil properties or plant characteristics, but also by the combined effects of soil and plant, as well as by the interactions of plant roots with the surrounding soil area. There are many extractants used to determine plant available Mn in soils; these extractants may be classified into five groups: a) water and dilute neutral salt solutions, b) ammonium acetate (pH 4 and 7), c) dilute acids (0.1N H_3PO_4 or double acid 0.1M HCl and 0.033M H_2SO_4 or Mehlich-3 reagent), d) chelate solutions, such as DTPA and EDTA [48].

There are contradictory results concerning the correlation of Mn uptake to soil extractable Mn determined by various extractants; some researchers found better correlations when chelate solutions (especially DTPA) were used, while some others obtained better results when acid solutions were used.

Mn Uptake by Plants

Manganese is basically taken up by plants as Mn^{2+}, or through some of its organic soluble complexes. In the second case, Mn uptake is enhanced by the existence of mycorrhiza [22]. Manganese uptake is metabolically controlled [1], which means that it is realised through a mechanism of energetic absorption, *i.e.*, with energy consumption. Clarkson (1988) found that the presence of 10μM dinitrophenol (DNP), an inhibitor of efficient energy (ATP) production, in a nutrient solution reduced by 80% the absorption of Mn by *Saccharum officinarum* plants, which proves that Mn uptake is realized through a mechanism of energetic absorption [25].

Mn Transport and Distribution within Plant Tissues

As previously mentioned, Mn is usually absorbed by plants in the form of Mn^{2+}. In some cases, Mn chelates (usually complexes with EDTA) may contribute to the mineral nutrition of higher plants. These complexes are taken up more slowly, than Mn^{2+} [25]. Generally, it is known that Mn is rapidly taken up and translocated within plants; it is likely that Mn is not binding to insoluble organic ligands, either in root tissue, or in xylem fluid [1]. In contrast to that, Mn was found to be relatively immobile in plants and is scarcely translocated in the phloem [5]. Under Mn deficiency conditions in soil plants may have extremely low Mn concentrations in young leaves, while they may have sufficient Mn concentrations in their mature (older) leaves. This happens because many plant species can accumulate a considerable quantity of Mn in their root system, or in their older shoots and leaves when they are grown under Mn sufficient conditions, and, therefore, when Mn deficiency occurs, they are able to remobilize part of this Mn quantity from these tissues to younger leaves in order to fulfill their nutritional needs in Mn [49].

Roles of Mn in Physiological and Biochemical Functions of Plants

Manganese plays a crucial role in many biochemical, as well as physiological operations of plants. Concerning biochemical operations, Mn is constituent of three enzyme complexes; of the complex of oxygen liberation of PSII, of the isoenzyme of MnSOD, and of the acid phosphatases complex [6]. The most important and unique physiological role of Mn appears to be its participation in the water splitting and O_2-evolving system of photosynthesis; it also plays a basic role in the photosynthetic electron transport system. Apparently, the Mn fraction that is loosely bound in chloroplasts is associated with O_2 evolution, whereas the firmly bound Mn fraction is involved in the electron pathway in photosynthesis [1]. Finally, the very firmly bound Mn in chloroplasts only contributes to their structural integrity [6]. The molecular mechanism of water oxidation in photosystem II still remains a mystery. The complex of oxygen evolution, which is a subunit of PSII and in which takes place the reaction of water oxidation, has in its molecule 4 atoms of Mn as cofactors, one of Ca and one of Cl. Recent advances in crystalography revealed the adherence of 7 aminoacids in the molecule of Mn_4Ca; however, its structural role is not clear yet [50, 51]. During water photolysis (splitting), two of its molecules are splitted towards $4H^+$, $4e^-$ and $2 O_2$. The four electrons, which are 'liberated' from water photolysis, are transferred to the enzyme of tyrosine, which then reduces the photosystem P680. P680 needs another external source of electrons in order to replenish the lack of e^- from the oxidized molecules of chlorophyll a, which have been excited by the solar radiation. This external source of electrons is the photosystem PSII through the electron transport chain, in which Mn participates with changes of its strength (Fig. **1**).

Superoxide dismutases is a group of metalloenzymes, widely distributed in biological systems, which catalyze the transformation of reactive oxygen species (ROS) into free oxygen and H_2O_2. ROS, which derive from the union of one electron with free oxygen (O_2), may cause lipid hyperoxidation, as well as damage in the cell membranes, aminoacids and nucleic acids, which may lead up to the plant death [6]. All the above-mentioned damages caused by the action of ROS are described with the term 'oxidative stress'.

Hydrogen hyperoxide in a second step is splitted into H_2O and free oxygen, with the aid of the enzymes guaiacol peroxidase, catalase, ascorbate peroxidase *etc.* [8, 53]. All the above-mentioned information is described with the following reactions:

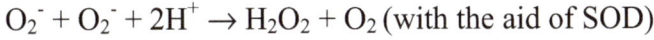

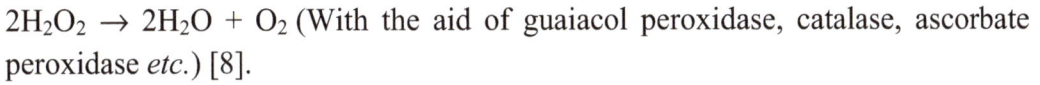

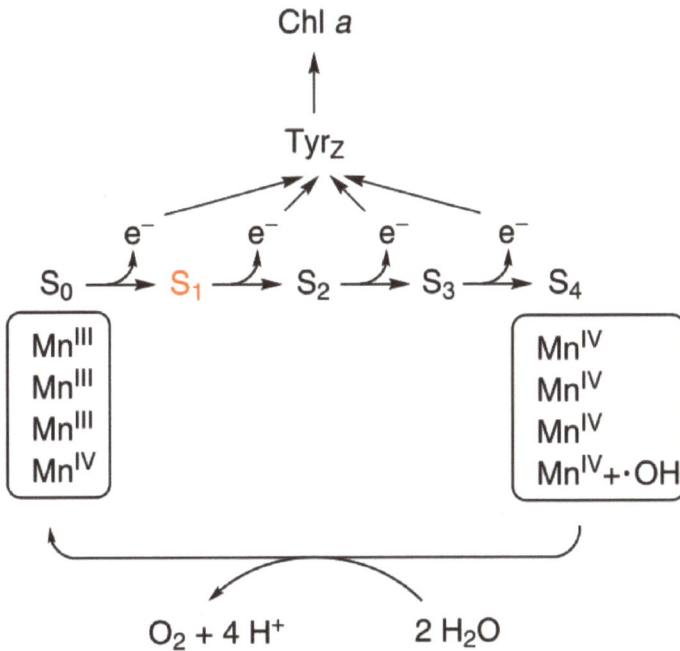

$$O_2 + e^- \rightarrow O_2^-$$

$$O_2^- + O_2^- + 2H^+ \rightarrow H_2O_2 + O_2 \text{ (with the aid of SOD)}$$

$$2H_2O_2 \rightarrow 2H_2O + O_2 \text{ (With the aid of guaiacol peroxidase, catalase, ascorbate peroxidase } etc.) \text{ [8]}.$$

Fig. (1). The role of Mn in the oxygen evolving complex (water splitting) (from: http://en.wikipedia.org/wiki/File:Kok-cycle.svg [52]).

There are 4 categories of SOD isoenzymes: a) isoenzyme of SOD with Mn (MnSOD), b) isoenzymes of SOD with Cu (CuSOD), c) with Zn (ZnSOD), d) isoenzymes with Fe (FeSOD). In wheat and tomato plants the isoenzymes of SOD with Cu and Zn have been found in chloroplasts, while the isoenzyme of MnSOD was found in mitochondria [54, 55]. In tobacco plants, MnSOD was found in chloroplasts [56]. Increased MnSOD activity, as well as activities of the other SOD isoenzymes under stress conditions have been referred by Mehlhorn and Wenzel (1996); Yu and Rengel (1999) and Shenker *et al.* (2004) [55, 57, 58].

Acid phosphatases are metalloenzymes found to contain Mn in their molecule. These enzymes, which catalyze the hydrolysis of phosphoric monoesters under acid conditions, have been isolated from a wide variety of vegetal species. The most studied and famous phosphatase is that found in *Ipomoea batatas* plants, with molecular weight 110KDa, containing 1-2 atoms of Mn in its molecule [6].

Manganese is an activator for about 35 enzymes, catalyzing reactions of oxidation-reduction, decarboxylations, hydrolysis reactions *etc.* In Table **2** are presented some of the most important enzymes, which are activated by Mn. Generally, Mn is involved in the metabolism of carbohydrates (reactions of glycolysis), of N, in the reactions of the Krebs cycle, in photosynthesis, in the biosynthetic way of shikimic acid, as well as in the activities of the enzymes IAA oxidase, polyphenol oxidase, allantoate amylohydrolase *etc.*

Glycolysis is a series of reactions during which glycose (which is produced from the split of polysaccharides) is splitted into two molecules of pyruvic acid, while it is also produced ATP and $NADH+H^+$. Totally 10 enzymes have been found to catalyze the reactions of glycolysis and 4 of them are activated by Mn; these enzymes are hexokinase, phosphoglycerine kinase, enolase and pyruvic kinase, in the first, 7th, 9th and 10th (last) step of glycolysis, respectively [6].

Referring to the influence of Mn on N metabolism, it should be pointed out that plants may utilize NH_4^+ or NO_3^- as source of N; however, nitric ions should be converted into NH_4^+ in order to be incorporated to plants' dry matter. That conversion (reduction) requires strong reducing agents. Reducing power is needed in order to supply the necessary e^-, so that the reduction of NO_3^- to NH_4^+ continues normally, without problems. Under Mn sufficieny, when photosynthetic rates are smooth, electrons are provided through the electrons transport chain. In contrast to that, under Mn starvation a decline in the electrons transport may take place, so NO_3^- are accumulated in plant tissues [1]. Arginase is an enzyme of the cycle of N that is activated by Mn; arginase splits arginine in urea and ornithine. Allantoate

amylohydrolase- which catalyzes the split of ureides (products of N metabolism)-
is another enzyme that has absolute demand for Mn in order to be activated [8].
Krebs cycle is a common metabolic pathway of oils, sugars and proteins. Totally,
9 reactions take place in this cycle and two enzymes catalyzing two reactions of
this cycle (isocitrate and malate dehydrogenase) are activated by Mn [8].

Table 2.   Some enzymes that are activated by Mn (modified from Burnell, 1988) [6].

NAD-malic enzyme
NADP-malic enzyme
NAD-isocitrate dehydrogenase
NADP- isocitrate dehydrogenase
Hydroxylamine reductase
Hexokinase
Phosphoglucokinase
Pyruvate kinase
UDP-glucose pyrophosphorylase
NAD kinase
Adenosine kinase
Arginine kinase
Phosphoglycomutase
Arginase
Alkaline phosphatase
Acid phosphatase
PEP carboxykinase
PEP carboxylase
Enolase, Glutamine synthetase, Pyruvate carboxylase, Peroxidase, Mn-Superoxide dismutase (MnSOD).

Manganese deficiency also negatively influences lipid biosynthesis. It has been
found that the content of the thylakoids of membranes in glycolipids and
unsaturated lipid acids was lower under Mn deficiency [8]. In seeds, Mn
starvation usually leads to lower oil, oleic acid, linoleic, palmitic, linolenic and
steatic acid contents. The reduced oil content of seeds may be probably ascribed
to the reduced photosynthetic rates of plants under Mn starvation (so, to the lack
of C skeletons for the biosynthesis of lipid acids) [7, 9, 59]. So, the influence of
Mn on lipid biosynthesis seems to be indirect; nevertheless, a direct influence of

Mn should not be excluded from possible reasons. Generally, the whole image concerning the influence of Mn on lipid acids' biosynthesis remains unclear and further research is needed in the near future in order to be clarified [8].

Concerning the photosynthesis of C4 plants, 3 enzymes have been found to be activated by Mn: PEP carboxylase, PEP carboxykinase and NAD-malic dehydrogenase; the first enzyme catalyzes the reaction of CO_2 to PEP (phosphoenolate form of pyrostafilic acid) and it may be also activated by Mg, while the second (catalyzing the decarboxylation of oxalate acid to PEP and CO_2) and the third one are only activated by Mn (they have absolute need from Mn in order to be activated). The absolute need for Mn for the photosynthesis of the C4 plants makes possible that these plant species are more susceptible, than the C3 ones, under Mn starvation [7, 8]. Many researchers found reduced photosynthetic rates of many plant species under Mn deficiency, than under Mn sufficiency [7, 9, 60].

There are many reports referring that under Mn starvation chlorophyll content in leaves is significantly reduced. The role of Mn on chlorophyll biosynthesis has not been fully clarified yet. It is possible that the reduced chlorophyll content is the consequence of hyperoxidosis and destruction of chloroplast membranes, or of the increased polyphenol oxidase activity [7]. In contrast to that explanation, Marschner (1995) supported that the changes in the structure of chloroplast membranes are owed to the impairment of lipid biosynthesis (because of the reduction of photosynthetic rates) and not to photooxidosis (hyperoxidation) of thylakoids and chlorophyll [8].

Apart from the above mentioned roles of Mn in basic physiological and biochemical reactions, Mn is also involved as cofactor in many other key-reactions, which lead to the biosynthesis of secondary metabolites of high importance for plants. The total of these key-reactions constitute the metabolic pathway of shikimic acid. A significant number of phenols are derived from intermediate metabolites of the biosynthetic pathway of shikimic acid. Among these phenols are included caffeic, chlorogenic, ferrulic acid *etc.*; their presence has been found to be correlated with increased resistance of plants to fungi attacks [6]. Huber and Wilhelm (1988) refer that young plants of *Cucurbita pepo* suffering from Mn deficiency were more prone to attacks from *Sclerotinia sclerotiorum*, than plants receiving adequate Mn [61]. During the biosynthetic pathway of shikimic acid, it is also produced tryptophan, an important aminoacid, which is a former substance of the biosynthesis of auxin (IAA). Tyrosine is another aminoacid that is produced during this biosynthetic pathway, while they are also produced coumarins and lignins [8]. According to Rengel *et al.* (1994), the impairment of the biosynthesis of lignin in the roots of *Triticum aestivum* plants due to the lack of Mn was the main reason for the reduced resistance of root system to attacks from the fungus *Gaeumannomyces graminis* var. *tritici* [62].

Soil and Non-Soil Conditions Favoring Mn Deficiency

There are soil and non-soil conditions favoring Mn deficiency in crops. The soil conditions, which may lead to Mn deficiency, are generally the following:

a)   **Deep organic soils formed on calcareous parent materials.**

b)   **Alluvial soils, rich in $CaCO_3$.**

c)   **Badly drained soils with great organic matter content.**

d)   **Very acid, highly leached, soils.**

e)   **Old acid soils on which organic matter and $CaCO_3$ added for many years.** Siebielec and Chaney (2006) limed some soils near a Ni refinery at Ontario in order to remediate them from Ni toxicity, but the repeated addition of limestone induced Mn deficiency in plants [63]. Overliming has been responsible for Mn deficiency in diverse regions of the world and it was identified 50 years ago as the main cause of Mn starvation for crops [64].

f)   **Soils with great moisture fluctuations [22].**

From the non-soil factors which may favorize Mn starvation, the most important is temperature. In Great Britain, Mn deficiency is more usual in cold regions, than in warmer ones. This may happen due to the fact that under low temperature conditions root growth is restricted. Later, when air and soil temperature increase, plants are capable of overcoming Mn deficiency, probably because of the greater root system growth. In addition to the above, Mn starvation may be observed in vegetal species with low reducing capacity of their root system, restricted ability of rhizosphere acidification, and/or low transpiration rates [64].

Critical Concentrations of Mn Deficiency and Symptoms

The concentration of Mn in plants at which growth is restricted by 10% is referred to as the critical Mn concentration [65]. Manganese deficiency level for most plant species is within the range from 10 to 20 mg/kg dry weight (d.w.) in mature leaves [5]. According to Kabata and Pendias (2001), the critical Mn deficiency level for most plant species ranges from 15 to 25 p.p.m. (d.w.), whereas the critical toxic Mn concentration is much more variable, depending on both plant and soil factors [1].

The most characteristic symptoms of Mn deficiency are the appearance of small yellow spots on the leaves and the interveinal chlorosis (Figs. **2-4**). In this respect that syndrome differs from that of Fe deficiency, where the whole young leaf becomes chlorotic (Fig. **5**). This kind of chlorosis in cases of Mn starvation happens due to decreased concentrations of chlorophyll (a and b) in leaves [66]. Chloroplasts are the most sensitive of all cell components to Mn deficiency and react by showing structural impairment. Changes in leaf chloroplast structure have been observed under conditions of Mn starvation [67]. According to Ohki (1985), the low chlorophyll content in leaves of plants suffering from Mn deficiency is the consequence of the limited number of photosynthetic active chloroplasts [60].

In *Spinacia oleracea* plants it was found that under Mn starvation the ratio chlorophyll a/b was 2.2, while in the control plants the relevant value was 2.8 [72]. In tobacco (*Nicotiana tabacum* L.) plants suffering from severe Mn deficiency, the levels of uronic acid (constituent of cell walls) in the root system were significantly reduced, while considerable changes in the content of pectins in arabinose, galactose and ramnose were also observed [73].

Fig. (2). In Mn deficiency the veins of the middle to upper leaves remain green, while the rest of the leaf becomes a uniform pale green to yellow (http://www.dpi.nsw.gov.au/agriculture/horti culture/greenhouse/pest-disease/general/cucumber-nutrition [68]).

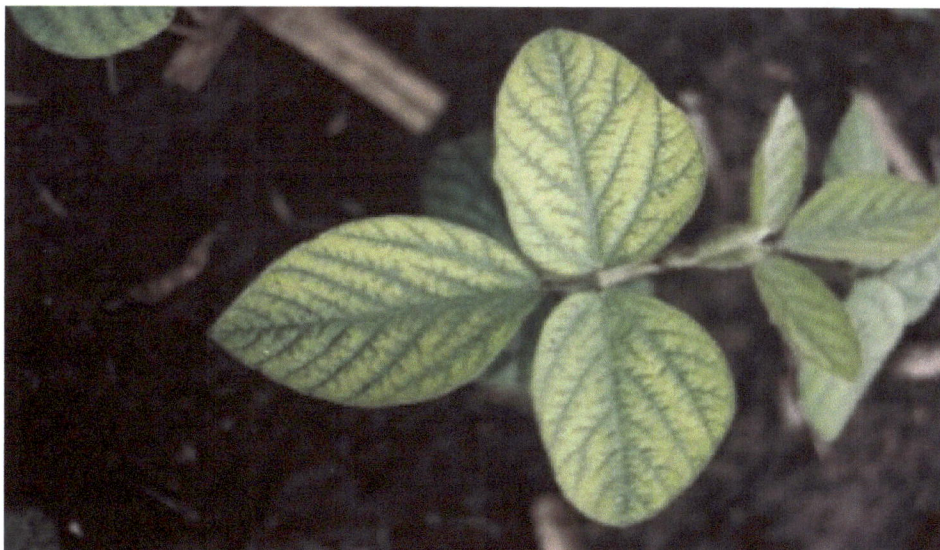

Fig. (3). Typical Mn deficiency symptoms in soybean plants (http://www.omafra.gov.on.ca/ english/crops/field/news/croppest/2007/12cpo07a2.htm) [69].

Fig. (4). Manganese deficient corn (http://hubcap.clemson.edu/~blpprt/acid2.html) [70].

Iron deficiency left, manganese deficiency right.
The chlorosis caused by iron deficiency is much more uniform as compared to the mosaic pattern of manganese deficiency, and with iron deficiency the veins change colour as well.

Fig. (5). Contrast between Fe (left) and Mn (right) deficiency in *Vitis vinifera* L. plants (http://www.omafra.gov.on.ca/IPM/english/grapes/plant-nutrition/iron.html) [71].

It has been found that Mn deficiency symptoms occur firstly in younger leaves as interveinal chlorosis, and at further stages necrotic spots on leaves and browning on roots appear. The appearance of Mn starvation symptoms firstly in young leaves is owed to the immobility of nutrient from the older leaves to the younger ones [65]. Oats have been found to be prone to Mn deficiency during the tillering stage. This kind of 'disease' is known as 'grey speck' [5]. Apart from oat, other sensitive to Mn deficiency crops are *Glycine max*, *Avena sativa*, *Triticum* sp, *Prunus persica*, *Malus domestica*, *Prunus cerasus*, *Citrus* sp., many species of the family *Cruciferae*, and some species of *Leguminosae* [8, 41].

Diagnosis of Mn Deficiency

The simplest and most economical method (since it does not demand technical support and laboratory equipment) to diagnose Mn deficiency is the observation of visual symptoms in plants. However, this is not a reliable way to predict and detect a nutrient imbalance in crops, since many times different nutrient deficiencies may be confused each other. In the case of Mn starvation, symptoms may be confused to those of Fe, Mg, or sometimes to those of S deficiency [65]. Another very important reason that the observation of deficiency symptoms alone should be avoided as detection method for nutritional disorders is the fact that, when the first symptoms appear, it is too late to correct the imbalance (plant metabolism has already been disturbed) [74].

The critical DTPA extractable Mn concentration in soils is 1 p.p.m. [75]. However, soil analysis alone is not a reliable method in diagnosing Mn deficiency, since Mn level in soils may be sufficient, but plants may suffer from Mn starvation. Although Mn constraint is relatively common in certain crops, grown on calcareous and alkaline soils, diagnosis and correction of the deficiency are not well defined. As soil analysis is not many times reliable alone in diagnosing Mn supply to plants, tissue tests should be considered together with soil results and field observations.

Foliar analysis is based on the collection of mature leaves from the middle of last vegetation in fruit trees. Composite leaf samples from different trees of the orchard are taken [75] in order to approach representative values of the 'real' nutritional status of Mn. Leaf Mn concentrations determined are compared to critical ones and the necessary fertilization (soil and/or foliar) recommendations follow, according to the results. The critical Mn concentrations depend on plant species; for example, for gramineous species critical Mn concentration is around 10-12 mg/kg d.w. but may vary from 10 to 50 mg/kg d.w. in other field crops and pasture species [65].

For a more acute determination of Mn starvation in plants, many techniques, such as enzyme analysis, chlorophyll fluorescence, gas exchange measurements *etc.* may be used in order to assess this nutritional disorder, since many times deficient concentrations of Mn in young leaves may be determined due to the immobility of Mn from mature to younger leaves. Chlorophyll a fluorescence has been used as a measure of photosynthetic dysfunction to permit early diagnosis of Mn deficiency; enhanced fluorescence at low Mn concentrations in leaves reflects a functional association between leaf Mn and electron transport from water to photosystem II [65]. Jiang *et al.* (2002) found that the maximum efficiency of PSII significantly decreased in Mn-starved maize leaves [76]. Finally, it has been found by many researchers that Mn starvation negatively affects the photosynthetic rates of many plant species [7-9, 60], but when Mn is resupplied photosynthesis quickly recovers in normal levels. However, the use of photosynthesis is not a reliable early diagnostic tool of Mn constraint, since when severe Mn deficiency occurs plant metabolism has been irreversibly disturbed.

Strategies Adopted by Plants in Order to Face Mn Starvation

There are many strategies adopted by plants in order to survive under Mn starvation. Some of these strategies are the acidification of rhizosphere, the remobilization of Mn from other tissues, having higher Mn concentrations (*e.g.*, root system), to younger leaves (usually suffering from low Mn concentrations), or from the older to the younger leaves, the formation of arbuscular mycorrhiza fungus (AMF) in the root system of plants in order to enhance Mn uptake, the increase of root exudates, the adjustment of root morphology to adverse soil conditions *etc.* All these strategies are fully analyzed below.

I.   **The acidification of rhizosphere and the reductive capacity of the root system.** Some vegetal species are able to acidify their rhizosphere under alkaline conditions (reduction of Mn^{4+} to Mn^{2+}, which can be absorbed by plants). Furthermore, differences in the reductive capacity of roots may affect rhizosphere pH and may explain genotypic differences in tolerance to Mn deficiency [64].

II.   **The remobilization of Mn from tissues having greater Mn concentrations (*e.g.*, root system, older leaves) to younger leaves, suffering from Mn deficiency.** According to Loneragan (1988), when plants grow into Mn deficiency from an adequate Mn supply, they may have extremely low Mn concentrations in their roots and young leaves, while maintaining high and adequate concentrations in their older leaves [49]. It is very characteristic the example of Nable and Loneragan (1984), who found that in subterranean clover Mn concentrations in root system, young and mature leaves under Mn deficient conditions were 13, 9 and 220 mg/kg d.w., respectively [77]. We found in our hydroponic experiments that Mn starved- for 130 days- olive plants had a very sufficient initial (before the beginning of the experiment) Mn quantity in their root system (95-100 mg/kg d.w.), that could be partially mobilized in order to satisfy the Mn nutritional needs of leaves (leaf Mn concentrations at the end of the experiment were about 20 mg/kg d.w.) [78]. It seems that in tree species the root system 'pool' of Mn which is stored under Mn sufficiency is a valuable nutrient source that can be remobilized under Mn-starved conditions in order to sufficiently satisfy the Mn nutritional needs of leaves.

III.   **The formation of arbuscular mycorrhiza fungus (AMF) in the root system of plants in order to enhance Mn uptake.** It is referred by many authors that AMF enhances Mn uptake under alkaline and/or Mn-deficient soil conditions [79, 80].

IV.   **The production of organic root exudates in order to facilitate Mn uptake through the formation of Mn-organic complexes.** The uptake of these complexes is usually favored by AMF [22]. According to Maruyama *et al.* (2005), who made a comparison of iron availability in leaves of barley and rice, the difference in Fe acquisition ability between these two species was found to be affected by differential mugineic acid secretion [81]. A similar possible mechanism might be responsible, according to Chatzistathis *et al.* (2009), for the differential Mn uptake between the olive cultivars 'Koroneiki' and 'Kothreiki' [47].

V.   **The adjustment of root morphology in order to be able to absorb more easily Mn^{2+}.** According to Rengel (2001), who made a review on genotypic differences in micronutrient use efficiency of many crops, micronutrient use efficient genotypes are capable of increasing plant available micronutrient pools in soil, through changing chemical and microbiological properties of the rhizosphere, as well as by growing thinner and longer roots and by having more efficient uptake and transport mechanisms [82]. We concluded from our experiments that the differential root system morphology of the olive cultivar 'Kothreiki' (this was richly-branched and with much greater root hair development and density), compared to those of the cultivars 'Koroneiki' and 'Chondrolia Chalkidikis', was probably responsible for the significantly greater Mn uptake [83].

Mn Use Efficient Genotypes

Under Mn deficiency, particular emphasis should be given to the production of genotypes with increased ability of Mn uptake and use efficiency [43, 45, 47, 64, 84-86]. Some of the mechanisms of Mn efficiency, based on which genotypes may be categorized into efficient and inefficient to Mn starvation are: i) genetic differences in internal requirements for Mn, ii) genetic differences in internal redistribution, iii) faster specific rate of uptake at low soil Mn concentrations, iv) better root geometry and v) greater root excretion of substances into the rhizosphere (genetic control of root exudates) to mobilize insoluble Mn (H$^+$, reductants, Mn-binding ligands, microbial stimulants) [41].

According to Jiang and Ireland (2005) and Jiang (2006), Mn efficient wheat cultivars own this ability to a better internal use of Mn, rather than to a higher Mn uptake and accumulation [87, 88]. In another study of Jiang, that of Jiang (2008), it has been found that Mn use efficiency of the wheat cultivar 'C8MM' has been related to a higher accumulation of Mn, while for the cultivars 'Paragon' and 'Maris Butler' the high internal use efficiency of Mn was highly related to an improved photosynthetic efficiency [89]. In addition to this, Chatzistathis *et al.* (2009) found that between the olive cultivars 'Koroneiki' and 'Kothreiki' the first one had better internal Mn use efficiency (better transport from root system to shoot), compared to the second one, so it was Mn-efficient, while 'Kothreiki' was Mn-inefficient [47]. So, it seems that in many cases the characteristic of Mn utilization efficiency is related more to a higher internal efficiency (transport from root system to shoot), rather than to a higher uptake and accumulation. Hebbern *et al.* (2005) found that Mn efficiency, based on grain yields, could not be related to whole shoot Mn concentration or Mn uptake by barley plants [90]. In contrast to that, in another experiment with barley plants, it was found that differential tolerance to low Mn availability has been related to differential capacity for high-affinity Mn uptake (Mn-efficient genotype 'Vanessa' had almost 4 times higher V_{max}, than the inefficient 'Antonia') [91].

According to Jhanji *et al.* (2012), growing Mn efficient genotypes, *i.e.*, those of producing high yields at low Mn supply, would represent a long-term solution and sustainable approach to crop production. From a practical point of view, genotypes that produce high grain yields at low level of Mn and respond well to Mn additions are the most desirable because they are able to express their high yield potential in a wide range of Mn availability [92]. In many cases, genotypes have been categorized into Mn efficient and inefficient, based on their ability to tolerate growth at low Mn concentrations. It has been found that the induction of Mn deficiency in two barley (*Hordeum vulgare* L.) genotypes, differing in Mn efficiency, led to a faster decline of the quantum yield efficiency in the Mn-inefficient genotype, than in the Mn-efficient one [93].

In a molecular basis, it has been found by Pedas *et al.* (2008) that under conditions of both Mn and Fe deficiency it has been induced an up-regulation of the gene HvIRT1 in two barley genotypes differing in Mn efficiency, but the expression levels in all cases were higher (up to 40%) in the Mn-efficient genotype [94].

Mn Fertilizers

Under Mn deficiency the use of fertilizers having the ability to oxidize the environment of rhizosphere (*e.g.*, ammonium sulphate) should be preferred. Two of these fertilizers are $(NH_4)_2SO_4$ and urea; when Mn availability is limited, NO_3^- fertilizers should be avoided for N fertilization, as they may deteriorate the problem of Mn starvation for plants. According to Siebelec and Chaney (2006), no severe Mn deficiency was observed when nitrogen was applied as combination of urea, ammonia and nitrates, than when applied only as nitrates in overlimed oat and red beet crops [63].

Many inorganic and organic sources of Mn have been tested for their suitability as Mn fertilizers. Their agronomic effectiveness is strongly influenced by their solubility in water, as well as by soil properties and the method of application. Manganous sulphate ($MnSO_4$) is most widely used both for soil applications, as well as in foliar sprays [64]. Manganese chelates, such as MnEDTA, is another source of Mn that can be used to alleviate Mn deficiency in crops, either as soil fertilizer, or as foliar spray. The problem with MnEDTA is that its instability in soil precludes its wider use. The chelated Mn can be readily displaced by Fe in acid soils, or by Ca in alkaline soils, thus Mn deficiency can be intensified [95-97].

On certain alkaline, calcareous and highly organic soils, applied Mn is rapidly immobilized and Mn deficiency can not be alleviated by soil applications. This is why foliar applications should be preferred under adverse soil conditions (when soil application is not suitable in correcting Mn deficiency). It has been found that foliar application is 10 times more effective than soil application because of the more uniform distribution pattern and the absence of immobilization [40]. The application of Mn in foliar sprays has been widely used until now in order to correct Mn starvation in annual crops, pastures and orchards. Several sources of Mn, including $MnSO_4$, MnO, $MnCl_2$ and Mn chelates, have been successfully used in foliar sprays [64]. In an experiment of Papadakis *et al.* (2005), it was found that $MnSO_4.H_2O$ was more effective than MnEDTA regarding the improvement of leaf Mn concentrations of 'Washington navel' orange trees when applied at equal Mn concentrations. Concerning the dose of Mn that should contain the foliar spray, it is referred by Papadakis *et al.* (2005), that the most appropriate was 800 or 1200 mg/l Mn (leaf Mn concentrations were found to be within the sufficiency range, *i.e.,* >25 mg/kg d.w.) [98]. In another study of Papadakis, that of Papadakis *et al.* (2008), it was found that when leaf Mn concentration was equal or higher to 14 mg/kg dry weight before the spray treatment, only one application of $MnSO_4.H_2O$ (800 mg/l Mn) on February was sufficient to satisfy the Mn nutritional needs of 'Washington navel' orange trees for a whole year [99]. According to Somasundaram *et al.* (2011), foliar application of 0.1% Mn sulphate may be effective in correcting Mn deficiency in *Citrus* trees [40].

Abadi *et al.* (2012) used municipal compost for fertilization and found that the effect of that application on leaf and root Mn in medicinal plants was significant [100]. Other organic amendments, such as animal manures, biosolids or composts have been reported to increase total amount of Mn in soils [101], thus they could be used as valuable source of Mn. According to Therios (1996), Mn content of the different kind of manures (from different animals) varies significantly among them [74]. Yolcu *et al.* (2011) used cattle manure together with zeolite and leonardite as organic fertilizers, and found an impressive effect on Mn content of ryegrass (*Lolium multiflorum* Lam.). According to the same authors, all these fertilizers (especially zeolite) showed great potential for use in organic agriculture [102]. Other organic materials, such as gyttja, alsil, humic acid, sea moss, straw and peat could be used as organic amendments, and it was found that the effect of these products on leaf Mn content of *Pistacia vera* trees was significant [103].

REFERENCES

[1]   Kabata-Pendias A, Pendias H. Trace Elements in Soils and Plants. 3rd ed. CRC Press, USA, 2001.
[2]   Gilkes RJ, McKenzie RM. Geochemistry and mineralogy of manganese in soils. In: Graham RD, Hannam RJ, Uren NC, Eds 'Manganese in soils and plants', Proceedings of the International symposium on 'Manganese in soils and plants', Kluwer Academic Publishers, 1988; pp. 23-35.
[3]   Lindsay WL. Chemical equilibria in soils 1979; pp.151-160.
[4]   McKenzie RM. Manganese oxides and hydroxides. In: Dixon JB, Weed SB, Eds 'Minerals in soil environments', Published by Soil Science Society of America, Madison, Wisconsin USA, 1977; pp. 181-193.

[5]   Mengel K, Kirkby E. Zinc. In: Mengel K, Kirkby E, Kosegarten H, Appel T, Eds. Principles of Plant Nutrition, 5th Edition. Kluwer Academic Publishers, Dordrecht, The Netherlands 2001; pp. 573-583.

[6]   Burnell JM. The biochemistry of manganese in plants. In: Graham RD, Hannam RJ, Uren NC, Eds 'Manganese in soils and plants', Proceedings of the International symposium on 'Manganese in soils and plants'. Kluwer Academic Publishers, Dordrecht, The Netherlands 1988; pp. 125-137.

[7]   Campbell LC, Nable RO. Physiological functions of manganese in plants. In: Graham RD, Hannam RJ, Uren NC, Eds 'Manganese in soils and plants', Proceedings of the International symposium on 'Manganese in soils and plants'. Kluwer Academic Publishers, Dordrecht, The Netherlands 1988; pp. 139-154.

[8]   Marschner H. Mineral Nutrition of Higher Plants. 2nd edition. Academic Press, London 1995; pp. 324-333.

[9]   Singh P, Misra A, Srivastava NK. Influence of Mn deficiency on growth, chlorophyll content, physiology and essential monoterpene oil(s) in genotypes of spearmint (*Mentha spicata* L.). Photosynthetica 2001; 39: 473-476.

[10]   Aubert H, Pinta M. Trace elements in soils. Elsevier Scientific Publishing Company, 1977; pp. 43-53.

[11]   Taylor RM, McKenzie RM. The association of trace elements with manganese minerals in Australian soils. Aust J Soil Res 1966; 4: 29-39.

[12]   Smith KA, Paterson JE. Manganese and Cobalt. In: Alloway BJ, Ed 'Heavy metals in soils'. Blackie Academic and Professional. 2nd ed. 1990; pp. 225-243.

[13]   Graven EH, Attoe OJ, Smith D. Effect of liming and flooding on manganese toxicity in alfalfa. Soil Sci Soc Am J 1965; 29: 702-706.

[14]   Mandal LN. Transformation of Iron and Manganese in water-logged rice soils. Soil Sci 1961; 91: 121-126.

[15]   Marschner H. Mechanisms of manganese acquisition by roots from soils. In: Graham RD, Hannam RJ, Uren NC, Eds 'Manganese in soils and plants', Proceedings of the International symposium on 'Manganese in soils and plants'. Kluwer Academic Publishers, Dordrecht, The Netherlands 1988; pp. 191-204.

[16]   Alifragis D. Soil: Genesis, Properties and Classification. Volume I. Aivazi Publications, Thessaloniki, Greece 2008; pp. 487-492. (In Greek).

[17]   Grass LB, MacKenzie AJ, Meek BD, Spencer WF. Manganese and Iron solubility changes as a factor in tile drain clogging: I. Observations during flooding and drying. Soil Sci Soc Am Proc 1973; 37: 14-17.

[18]   Patrick WF, Turner FT. Effect of redox potential on manganese transformation in waterlogged soil. Nature 1968; 220: 476-478.

[19]   Gotoh S, Patrick WH. Transformation of manganese in a waterlogged soil as affected by redox potential and pH. Soil Sci Soc Am Proc 1972; 36: 738-741.

[20]   Wei X, Hao M, Shao M, Gale WJ. Changes in soil properties and the availability of soil micronutrients after 18 years of cropping and fertilization. Soil Til Res 2006; 91: 120-130.

[21]   El-Jaoual T, Cox DA. Manganese toxicity in plants. J Plant Nutr 1998; 21: 353-386.

[22]   Alifragis D, Papamichos N. Fertility of forest soils; 1994. Dedousis Publications, Thessaloniki, Greece. (In Greek).

[23]   Douka AE. Oxidation of Mn^{2+} through the action of two Mn-oxidazing bacteria. Taxonomy and study of their morphology and physiology. Doctoral Thesis, National and Kapodestrian University of Athens, 1973.

[24]   Ghiorse WC. The biology of manganese transforming microorganisms in soil. In: Graham RD, Hannam RJ, Uren NC, Eds 'Manganese in soils and plants', Proceedings of the International symposium on 'Manganese in soils and plants'. Kluwer Academic Publishers, Dordrecht, The Netherlands 1988; pp. 75-85.

[25]   Clarkson DT. The uptake and translocation of manganese by plant roots. In: Graham RD, Hannam RJ, Uren NC, Eds 'Manganese in soils and plants', Proceedings of the International symposium on 'Manganese in soils and plants'. Kluwer Academic Publishers, Dordrecht, The Netherlands 1988; pp. 101-111.

[26]   Chatzistathis T. Investigation of the role of Mn on olive trees' mineral nutrition. Doctoral Thesis, 2008. Aristotle University of Thessaloniki, Greece. (In Greek).

[27]   Ghasemi-Fasaei R, Ronaghi A, Maftoun M, Karimian N, Soltanpour PN. Influence of FeEDDHA on Iron-Manganese Interaction in Soybean Genotypes in a Calcareous Soil. J Plant Nutr 2002; 26: 1815-1823.

[28]   Quartin VML, Antunes ML, Muralha MC, Sousa MM, Nunes MA. Mineral imbalance due to manganese excess in Triticales. J Plant Nutr 2001; 24: 175-189.

[29]   Mousavi SR, Shahsavari M, Rezaei M. A general overview on Mn importance for crops production (Review). Aus J Basic Appl Sci 2011; 5: 1799-1803.

[30]   Monnet F, Vaillant N, Vernay P, Coudret A, Sallanon H, Hitmi A. Relationship between PSII activity, CO_2 fixation, and Zn, Mn and Mg contents of *Lolium perenne* under zinc stress. J Plant Physiol 2001; 158: 1137-1144.

[31]   Nautiyal N, Chatterjee C. Copper-Manganese interaction in cauliflower. J Plant Nutr 2002; 25: 1701-1707.

[32]   Aref F. Manganese, iron and copper contents in leaves of maize plants (*Zea mays* L.) grown with different boron and zinc micronutrients. Afr J Biotech 2012; 11: 896-903.

[33]   Grasmanis VO, Leeper GW. Toxic manganese in near-neutral soils. Plant and Soil 1966; XXV(1): 41-48.

[34]   Cotter DJ, Mishra UN. The role of organic matter in soil manganese equilibrium. Plant and Soil 1968; XXIX(3): 439-448.

[35]   Boyd HW. Manganese toxicity to peanuts in autoclaved soil. Plant Soil 1971; 34: 133-144.

[36]   Bromfield SM. The properties of a biologically formed manganese oxide, its availability to oats and its solution by root washings. Plant and Soil 1958; IX(4): 325-337.

[37]   Posta K, Marschner H, Romheld V. Manganese reduction in the rhizosphere of mycorrhizal and nonmycorrhizal maize. Mycorrhiza 1994; 5: 119-124.

[38]   Kothari SK, Marschner H, Romheld V. Effect of a vesicular arbuscular mycorrhizal fungus and rhizosphere microorganisms on manganese reduction in the rhizosphere and manganese concentrations in maize (*Zea mays L.*). New Phytol 1991; 117: 649-655.

[39]   Tong Y, Rengel Z, Graham RD. Interactions between nitrogen and manganese nutrition of barley genotypes differing in manganese efficiency. Ann Bot 1997; 79: 53-58.

[40]   Somasundaram J, Meena HR, Singh RK, Prasad SN, Parandiyal AK. Diagnosis of micronutrient imbalance in lime crop in semi-arid region of Rajasthan, India. Com Soil Sci Plant Anal 2011; 42: 858-869.

[41]   Graham RD. Genotypic differences in tolerance to manganese deficiency. In: Graham RD, Hannam RJ, Uren NC. 'Manganese in soils and plants', Proceedings of the International

symposium on 'Manganese in soils and plants'. Kluwer Academic Publishers, Dordrecht, The Netherlands 1988; pp. 261-276.

[42]   Foy CD, Weil RR, Coradetti CA. Differential manganese tolerances of cotton genotypes in nutrient solution. J Plant Nutr 1995; 18: 685-706.

[43]   Khabaz-Saberi H, Graham RD, Rathjen AJ. Genotypic variation for Mn efficiency in durum wheat (*Triticum turgidium* L. var. *durum*). Plant Nutrition-for sustainable food production and environment; 1997: 289-290.

[44]   Rout GR, Samantaray S, Das P. Studies on differential manganese tolerance of mung bean and rice genotypes in hydroponic culture. Agronomie 2001; 21: 725-733.

[45]   Sadana US, Lata K, Claassen N. Manganese efficiency of wheat cultivars as related to root growth and internal manganese requirement. J Plant Nutr 2002; 25: 2677-2688.

[46]   Gherardi MJ, Rengel Z. Genotypes of lucerne (*Medicago sativa* L.) show differential tolerance to manganese deficiency and toxicity when grown in bauxite residue sand. Plant Soil 2003; 249: 287-296.

[47]   Chatzistathis T, Therios I, Alifragis D. Differential uptake, distribution within tissues, and use efficiency of Manganese, Iron and Zinc by olive cultivars 'Kothreiki' and 'Koroneiki'. HortSci 2009; 44: 1994-1999.

[48]   Reisenauer H.M. Determination of plant-available soil manganese. In: Graham RD, Hannam RJ, Uren NC, Eds 'Manganese in soils and plants', Proceedings of the International symposium on 'Manganese in soils and plants'. Kluwer Academic Publishers, Dordrecht, The Netherlands 1988; pp. 87-98.

[49]   Loneragan JF. Distribution and movement of manganese in plants. In: Graham RD, Hannam RJ, Uren NC, Eds 'Manganese in soils and plants', Proceedings of the International symposium on 'Manganese in soils and plants'. Kluwer Academic Publishers, Dordrecht, The Netherlands 1988; pp. 113-124.

[50]   Kusunoki M. Mono-manganese mechanism of the photosystem II water splitting reaction by a unique Mn_4Ca cluster. Biochimica et Biophysica Acta 2007; 1767: 484-492.

[51]   Nagata T, Nagasawa T, Zharmukhamedov SK, Klimov VV, Allakhverdiev SI. Reconstitution of the water-oxidizing complex in manganese-depleted photosystem II preparations using synthetic binuclear Mn(II) and Mn (IV) complexes: production of hydrogen peroxide. Photos Res 2007; 93: 133-138.

[52]   Oxygen-evolving complex. From Wikipedia, the free encyclopedia: http://en.wikipedia.org/ wiki/File:Kok-cycle.svg

[53]   Lidon FC, Teixeira MG. Oxy radicals production and control in the chloroplast of Mn-treated rice. Plant Sci 2000; 152: 7-15.

[54]   Wu G, Wilen RW, Robertson AJ, Gusta LV. Isolation, chromosomal localization, and differential expression of mitochondrial manganese superoxide dismutase and chloroplastic copper/zinc superoxide dismutase genes in wheat. Plant Physiol 1999; 120: 513-520.

[55]   Shenker M, Plessner OE, Tel-Or E. Manganese nutrition effects on tomato growth, chlorophyll concentration, and superoxide dismutase activity. J Plant Physiol 2004; 161: 197-202.

[56]   Yu Q, Osborne LD, Rengel Z. Increased tolerance to Mn deficiency in transgenic tobacco overproducing superoxide dismutase. Ann Bot 1999; 84: 543-547.

[57]   Mehlhorn H, Wenzel A. Manganese deficiency enhances ozone toxicity in bush beans (*Phaseolus vulgaris* L. cv. Saxa). J Plant Physiol 1996; 148: 155-159.

[58]   Yu Q, Rengel Z. Micronutrient deficiency influences plant growth and activities of Superoxide Dismutases in narrow-leafed lupins. Ann Bot 1999; 83: 175-182.

[59]   Wilson DO, Boswell FC, Ohki K, Parker MB, Shuman LM, Jellum MD, Changes in soybean seed oil and protein as influenced by manganese nutrition. Crop Sci 1982; 22: 948-952.

[60]   Okhi K. Manganese deficiency and toxicity effects on photosynthesis, chlorophyll, and transpiration in wheat. Crop Sci 1985; 25: 187-191.

[61]   Huber DM, Wilhelm NS. The role of manganese in resistance to plant diseases. In: Graham RD, Hannam RJ, Uren NC, Eds 'Manganese in soils and plants', Proceedings of the International symposium on 'Manganese in soils and plants'. Kluwer Academic Publishers, Dordrecht, The Netherlands 1988; pp. 155-173.

[62]   Rengel Z, Graham RD, Pedler J. Time-course of biosynthesis of phenolics and lignin in roots of wheat genotypes differing in manganese efficiency and resistance to take-all fungus. Ann Bot 1994; 74: 471-477.

[63]   Siebielec G, Chaney RL. Manganese fertilizer requirement to prevent manganese deficiency when liming to remediate Ni-phytotoxic soils. Com Soil Sci Plant Anal 2006; 37: 163-179.

[64]   Reuter DJ, Alston AM, McFarlane JD. Occurrence and correction of manganese deficiency in plants. In: Graham RD, Hannam RJ, Uren NC, Eds 'Manganese in soils and plants', Proceedings of the International symposium on 'Manganese in soils and plants'. Kluwer Academic Publishers, Dordrecht, The Netherlands 1988; pp. 205-224.

[65]   Hannam RJ, Ohki K. 1988. Detection of manganese deficiency and toxicity in plants. In: Graham RD, Hannam RJ, Uren NC, Eds 'Manganese in soils and plants', Proceedings of the International symposium on 'Manganese in soils and plants'. Kluwer Academic Publishers, Dordrecht, The Netherlands 1988; pp. 243-259.

[66]   Shenker M, Plessner OE, Tel-Or E. Manganese nutrition effects on tomato growth, chlorophyll concentration, and superoxide dismutase activity. J Plant Physiol 2004; 161: 197-202.

[67]   Papadakis IE, Bosabalidis AM, Sotiropoulos TE, Therios IN. Leaf anatomy and chloroplast ultrastructure of Mn-deficient orange plants. Acta Physiol Plant 2007; 29: 297-301.

[68]   http://www.dpi.nsw.gov.au/agriculture/horticulture/greenhouse/pest-disease/general/cucumber-nutrition. Nutrient disorders of greenhouse Lebanese cucumbers.

[69]   http://www.omafra.gov.on.ca/english/crops/field/news/croppest/2007/12cpo07a2.htm Manganese deficiency. Ontario: Ministry of Agriculture and Food.

[70]   http://hubcap.clemson.edu/~blpprt/acid2.html. Soil acidity and liming. Internet inservice training.

[71]   http://www.omafra.gov.on.ca/IPM/english/grapes/plant-nutrition/iron.html. Iron deficiency. Ontario: Ministry of Agriculture, Food and Rural Affairs. Ontario GrapeIPM.

[72]   Anderson JM, Pyliotis NA. Studies with manganese-deficient spinach chloroplasts. Biochimica et Biophysica Acta (BBA)-Bioenergetics 1969; 189: 280-293.

[73]   Wang JJ, Evangelou BP, Ashraf MM. Changes in root cell wall chemistry induced by manganese exposure with two tobacco genotypes. J Plant Nutr 2003; 26: 1527-1540.

[74]   Therios I. Plant mineral nutrition and fertilizers 1996; Dedousis Publications, Thessaloniki, Greece. (In Greek).

[75]   Lindsay WL, Norvell WA. Development of a DTPA soil test for zinc, iron, manganese and copper. Soil Sci Soc Am J 1978; 42: 6421-6428.

[76]   Jiang CD, Gao HY, Zou Q. Characteristics of photosynthetic apparatus in Mn-starved maize plants. Photosynthetica 2002; 40: 209-213.

[77]   Nable RO, Loneragan JF. Translocation of Mn in subterranean clover (Trifolium subterraneum L., cv. Seaton Park). I. Redistribution during vegetative growth. Aus J Plant Physiol 1984; 11: 101-111.

[78]   Chatzistathis T, Papadakis I, Therios I, Giannakoula A, Dimassi K. Is chlorophyll fluorescence technique a useful tool to assess Manganese deficiency and toxicity stress in olive plants? J Plant Nutr 2011; 34: 98-114.

[79]   Wu QS, Li GH, Zou YN. Roles of arbuscular mycorrhizal fungi on growth and nutrient acquisition of peach (*Prunus persica L. Batsch*) seedlings. J Anim Plant Sci 2011; 21: 746-750.

[80]   Sohn BK, Kim KY, Chung SJ, *et al.* Effect of different timing on AMF inoculation on plant growth and flower quality of chrysanthemum. Sci Hortic 2003; 98: 173-183.

[81]   Maruyama, T, Higuchi, K, Yoshida M, Tadano, T. Comparison of iron availability in leaves of barley and rice. Soil Sci Plant Nutr 2005; 51: 1037-1042.

[82]   Rengel Z. Genotypic differences in micronutrient use efficiency in crops. Com Soil Sci Plant Anal 2001; 32: 1163-1186.

[83]   Chatzistathis T, Orfanoudakis M, Alifragis D, Therios I. Colonization of Greek olive cultivars' root system by arbuscular mycorrhiza fungus: root morphology, growth and mineral nutrition of olive plants. Sci Agric 2013; 70: 185-194.

[84]   Huang C, Graham RD. Efficient Mn uptake in barley is a constitutive system. Plant nutrition-for sustainable food production and environment 1997; 269-270.

[85]   Pearson JN, Rengel Z. Genotypic differences in the production of carbohydrates between roots and shoots of wheat, grown under zinc or manganese deficiency. Ann Bot 1997; 80: 803-808.

[86]   Sadana US, Samal D, Claassen N. Differences in manganese efficiency of wheat (*Triticum aestivum* L.) and raya (*Brassica juncea* L.) as related to root-shoot relations and manganese influx. J Plant Nutr Soil Sci 2003; 166: 385-389.

[87]   Jiang WZ, Ireland CR. Characterization of manganese use efficiency in UK wheat cultivars grown in a solution culture system and in the field. J Agric Sci 2005; 143: 151-160.

[88]   Jiang WZ. Manganese use efficiency in different wheat cultivars. Env Exp Bot 2006; 57: 41-50.

[89]   Jiang WZ. Comparison of responses to Mn deficiency between the UK wheat genotypes Maris Butler, Paragon and the Australian wheat genotype C8MM. J Integr Plant Biol 2008; 50: 457-465.

[90]   Hebbern CA, Pedas P, Schjoerring JK, Knudsen L, Justed S. Genotypic differences in manganese efficiency: Field experiments with winter barley (*Hordeum vulgare* L.). Plant Soil 2005; 272: 233-244.

[91]   Pedas P, Hebbern CA, Schjoerring JK, Holm PE, Husted S. Differential capacity for high-affinity Mn uptake contributes to differences between barley genotypes in tolerance to low Mn availability. Plant Physiol 2005; 139: 1411-1420.

[92]   Jhanji S, Sadana US, Sekhon NK, Gill TPS, Khurana MPS, Kaur R. Screening diverse rice (Oryza sativa L.) genotypes for Mn efficiency. Proceedings of the National Academy of Sciences, India, Section B-Biological Sciences 2012; 82: 447-452.

[93]   Husted S, Laursen KH, Hebbern CA, Schmidt SB, Pedas B, Haldrup A, Jensen PE. Manganese deficiency leads to to genotype-specific changes in fluorescence induction kinetics and state transitions. Plant Physiol 2009; 150: 825-833.

[94]   Pedas P, Ytting CK, Fuglsang AT, Jahn TP, Schjoerring JK, Husted S. Manganese efficiency in barley: Identification and characterization of the metal ion transporter HvIRT1. Plant Physiol 2008; 148: 455-466.

[95]   Boxma R, DeGroot AJ. Behaviour of iron and manganese chelates in calcareous soils and their effectiveness for plants. Plant Soil 1971; 34: 741-749.

[96]   Knezek BD, Greinert H. Influence of soil Fe and MnEDTA interactions upon the Fe and Mn nutrition of bean plants. Agron J 1971; 63: 617-619.

[97]   Ryan J, Harig SN. Transformation of incubated micronutrient chelates in calcareous soils. Soil Sci Soc Am J 1983; 47: 806-810.

[98]   Papadakis IE, Protopapadakis E, Therios IN, Tsirakoglou V. Foliar treatment of Mn deficient 'Washington navel' orange trees with two Mn sources. Sci Hortic 2005; 106: 70-75.

[99]   Papadakis IE, Chatzissavvidis C, Therios IN, Protopapadakis EE. Changes in leaf Mn concentrations in mature 'Washington navel' orange trees with different Mn status after a single foliar application of MnSO$_4$. Eur J Hortic Sci 2008; 73: 162-165.

[100] Abadi ZA, Ghajar-Sepanlou M, Bahmanyar MA. The effect of municipal compost application on the amount of microelements and their absorption in soil and medicinal plant of mint (Menthas). Afr J Biotech 2012; 10: 17716-17725.

[101] He ZL, Yang XE, Stoffella PJ. Trace elements in agroecosystems and impacts on the environment. J Trace Elem Med Biol 2005; 19: 125-140.

[102] Yolcu H, Seker H, Gullap MK, Lithourgidis A, Gunes A. Application of cattle manure, zeolite and leonardite improves hay yield and quality of annual ryegrass (*Lolium multiflorum* Lam.) under semiarid conditions. Aus J Crop Sci 2011; 5: 926-931.

[103] Demirkiran AR, Cengiz MC. Effects of different organic materials and chemical fertilizers on nutrition of pistachio (*Pistacia vera* L.) in organic arboriculture. Afr J Biotech 2010; 9: 6320-6328.

CHAPTER 6

Copper Deficiency

Abstract: Copper is needed in very low concentrations by plants. The critical Cu deficiency level for adequate plant growth is around 2-4 p.p.m., depending on plants species and genotypes. Copper is involved in many physiological and biochemical functions, such as photosynthesis, electron transport in photosystem II (PSII), chloroplast ultrastructure, carbohydrate and nitrogen metabolism, water permeability of xylem vessels, as well as in the production of DNA and RNA; it is also related with mechanisms of disease resistance. The critical Cu concentration, according to the DTPA solution, is 0.2 p.p.m. Soil available concentrations of Cu depend on parent material, pH and $CaCO_3$ content, organic matter, phosphoric ion content, cation exchange capacity (C.E.C.), soil type, structure and moisture, the availability of other nutrients *etc.* Some of the most characteristic symptoms of Cu deficiency include the formation of narrow and twisted leaves, as well as leaf curling, while their petioles bend downward. Enhanced remobilization and retranslocation in order to overcome Cu starvation and survive are within the most important mechanisms of tolerance adopted by plants. Increased antioxidant defence mechanisms under Cu stress conditions include enzymatic and non-enzymatic antioxidant responses. When crops suffer from Cu deficiency, both soil or foliar applications may be a good solution; $CuSO_45H_2O$ is usually the substance that is preferred. Apart from Cu sulphate (25% Cu), other Cu sources that could be used as Cu fertilizers are: cupric oxychloride (50% Cu), cuprous oxide (32.8% Cu) and chelated Cu (5% Cu).

All these topics concerning Cu availability in soils and uptake by plants, the roles of Cu in plant metabolism, the critical concentrations and symptoms of Cu deficiency, as well as the tolerance mechanisms and strategies adopted by plants in order to survive under Cu starvation, and the Cu-fertilizers that can be used to overcome Cu deficiency in crops, are fully discussed and analyzed.

Keywords: COPT proteins, Cu availability, Cu deficiency, Cu fertilizers, Cu homeostasis, Cu sulphate, Cu tolerance , Cu uptake, cupric oxychloride.

INTRODUCTION

Copper deficiency is a sparse nutritional disorder for many crops [1, 2], since plant demand in Cu is low and most soils contain adequate levels of Cu to meet this requirement [3]. Copper content greatly varies among soil types, depending on parent material, as well as on the climatic conditions (enhanced leaching of Cu from the upper to the deeper soil layers of sandy soils in rainy areas) of each geographical area. The mean values for Cu vary from 13 to 24 p.p.m., being highest for kastanozems and chernozems and lowest for podzols and histosols [4]. The available for plant growth Cu concentrations in soils are influenced by pH, carbonate content, cation exchange capacity (C.E.C.), clay and organic matter content, soil humidity, the amount of Al, Fe and Mn oxides and hydroxides, the content of phosphate ions, the addition of organic amendments *etc.* Plant available Cu concentration in soils is influenced by the choice of the extractant used; greater Cu concentrations are usually determined by the choice of strong acids as extractants, than by chelating agents, such as DTPA. With the use of DTPA the critical Cu concentration was found to be 0.2 mg/kg soil, while much greater is the critical Cu concentration when strong acids, like HNO_3, are used. According to Ozyazici *et al.* (2011) Cu content in soils of Eastern Black Sea region, in Turkey, varied from 0.02 to 14.69 mg/kg soil [5].

Copper is contained in plastocyanin, an acidic protein containing 2 Cu atoms per molecule, as well as in the isoenzyme of superoxide dismutase with Cu (CuSOD). Generally, Cu occurs in enzymes having vital functions in plant metabolism. Furthermore, Cu influences carbohydrate and nitrogen metabolism, water permeability of xylem vessels, DNA and RNA production, while it is also involved in the mechanisms of disease resistance (impairement of lignification) [3, 4].

Critical Cu deficiency level depends on plant species; however, Cu levels less than 2 p.p.m. in tissues are likely to be inadequate for the growth of most plant species [4]. According to Mengel and Kirkby (2001), Cu is taken up by the plants in very low quantities, since Cu concentration in most plant species is low and varies within the range from 5 to 20 µg/g. d.w. [3]. Some of the most characteristic symptoms of Cu starvation are the formation of narrow and twisted leaves, leaf curling, as well as the inhibition of lignin synthesis. In order to detect and diagnose Cu deficiency, there are many methods, such as field observations (visual symptoms), soil testing, tissue sampling and analysis, as well as biochemical assays.

The purposes of this chapter are: i) to present the behaviour of Cu in soil profiles and its reaction with soil components, ii) to discuss how the different soil-and other-factors influence Cu availability and uptake, iii) to present and describe the advantages and disadvantages of the methods used to diagnose and detect Cu starvation, iv) to provide all the necessary information concerning Cu requirements of different crops, as well as to present all the relevant references on the critical deficiency concentrations of many plant species and on the Cu-containing fertilizers, applied in soil or in foliar.

Copper in Soils

Copper occurs in soils almost exclusively in divalent form. The largest fraction of Cu is usually present in the crystal lattices of primary and secondary minerals (non-available for plants Cu). In addition, a high proportion of Cu is bound by the negative charges of soil organic matter. Copper is adsorbed to inorganic and organic negatively charged groups and is dissolved in the soil solution as Cu^{2+} and organic Cu complexes [4]. The critical soil Cu concentration depends on the extractant solution used for the determination of plant available Cu; DTPA is the mostly used extractant for the determination of extractable micronutrient concentrations. Other extractants that can be used are $CaCl_2$, EDTA, Mehlich-1, Mehlich-3 *etc.* The acid solutions usually extract greater Cu concentrations than the other ones used, like salts and organic extractants. Indeed, Sequeira *et al.* (2011) found that Mehlich-1 and Mehlich-3 extracted greater quantity of of Cu, compared to DTPA, but foliar Cu content was significantly and positively correlated with soil Cu extracted by all the three solutions tested [6]. The critical Cu concentration, according to the DTPA solution, is 0.2 p.p.m. [7]. In contrast to that, Irmak (2010) refers that the critical Cu level in soils is 1 mg/kg [8]. Hitsuda *et al.* (2010) used Mehlich-1 as micronutrient extractant solution and found that the critical Cu concentration in marginal soils (where micronutrient deficiencies are likely to occur) was 0.2 mg/kg soil. According to the same authors, the determination method can clearly identify the critical values for micronutrient deficiency in soils [9]. It was found that the minimum and maximum Cu concentrations from surface soil samples received from the region of Elassona, Central Greece, were 0.14 and 1.78 mg/kg soil, respectively [10]. Furthermore, the DTPA extractable Cu concentration determined in limed surface soils from the semi-arid region of Rajastan, India, varied within the range from 0.14 to 0.46 mg/kg soil [11].

Copper sufficiency in soil does not necessarily guarantee the sufficiency for crops (there are soil factors limiting Cu availability for plants). According to Turan *et al.* (2013), Cu deficiency was found only in leaf and not in soil samples, in a survey conducted in olive groves in the provinces of Izmir, Manisa, Aydin, Balikesir, Canakkalea and Bursa, in the West Anatolia region, Turkey [12]. Edwards *et al.* (2012) found that about 25% of the Scottish soil agricultural samples measured were Cu-deficient. According to the same authors, a predicted risk assessment using soil parent material, texture and drainage status suggested that 22% of the agricultural area of Scotland was at high risk of Cu deficiency, while 38% of that area was at medium risk [13]. In a similar result concluded the research of Sinclair and Edwards (2008), who found that Cu deficiency is widespread and suggested to occur over 30, 25, 20 and 5% of the land area of Scotland, Germany, Finland and England and Wales, respectively [14]. In the study of Karyotis *et al.* (2011) it has been found that 45.8% of the soil samples received from hilly and mountainous areas of the region of Elassona, Central Greece, were below the deficiency limit [10]. According to Vogel and Jokela (2011), extremely low Cu concentrations in pine stands of the southeastern United States caused a serious decline in forest volume production [15].

Copper availability in soils depends on many factors, such as pH, organic matter, $CaCO_3$ and phosphoric ion contents, cation exchange capacity (C.E.C.), soil type and structure, soil moisture, the availability of other nutrients *etc.* The influence of each of these factors on Cu availability is shortly described below.

a.   **Organic matter.** Organic and peaty soils, as well as soils having moderate to high organic matter content, have been found to be most commonly deficient in Cu [16]. According to Pandian *et al.* (2011) the two factors that regulated Cu content of rice grains were soil organic carbon content and soil Cu availability [17]. In contrast to Alifragis (2008), in the study of Sharma *et al.* (2009) it has been found that DTPA extractable Cu increased with the increase of organic carbon [18]. It seems that in alkaline and calcareous soils (as those existed in the soil area in study of Sharma *et al.*, where pH values were greater than 8, reaching up to 9.5) organic Cu complexes are soluble and significantly contribute to satisfy the Cu nutritional needs of crops.

b.   **Parent material.** Parent material greatly influences Cu content of soils; Cu deficiency is much more likely to occur in acid soils, derived from siliceous parent material [13].

c.   **Soil type and texture.** According to Sequeira *et al.* (2011), Cu deficiency is common in red and black clayey soils, rich in Mn and Fe oxides [6]. Concerning the influence of texture on Cu content, it has been found that total Cu content was higher in the fine-textured, than in the coarse-textured soils. Total Cu contents increased with the increase of clay and decreased with the increase of sand content [18].

d.   **pH and carbonate content.** Generally, as pH increases, Cu solubility and availability are significantly decreased. It is much more likely to occur Cu starvation in alkaline soils, derived from carbonate- rich sediment [13].

e.   **The phosphoric ion content.** Generally, as for the other micronutrients (Fe, Mn, Zn), in soils having high P content the formation of insoluble phosphoric substances with Cu limit its' availability for plants [16]. This is why in heavy P-fertilized soils Cu availability may be decreased. It has been found that P application could effectively decrease the bioavailability of Cu in *Zea mays* L. and *Glycine max* L. crops [19]. However, in the study of Richards *et al.* (2011), it was found that the long-term application of inorganic P had little effect on Cu availability [20].

f.   **Soil moisture and drainage.** Under flooded, reducing conditions, Cu^{2+} may be converted to Cu^+, so Cu availability for plants is decreased, since plants absorb Cu as divalent cation.

g.   **Soil erosion.** Soil erosion decreases Cu content and uptake by plants. It was found that AB-DTPA extractable Cu concentrations were decreased due to the severity of erosion in surface and subsurface soils of the Sharkul area of Pakistan. In contrast to that, in the substrata soils (60-75 cm depth) the effect of erosion was almost non-significant [21].

h.   **The availability of other nutrients.** In many cases, Cu concentrations in soils may be sufficient, but plants may suffer from Cu deficiency due to the interaction effect of other nutrients on Cu uptake. According to Yuan-Rong *et al.* (2011), the Guizhou semi-fine sheep in the Weining County, Guizhou province, south west China, was found to suffer from Cu starvation, but this nutritional disorder was caused by secondary Cu deficiency (due to high S and Mo), since soil and forage Cu concentrations were within the normal ranges [22]. It has been found that N application could significantly increase maize shoot Cu concentration and translocation from soil to the above ground tissues [19].

Copper Uptake, Mobilization and Distribution

Copper is absorbed as Cu^{2+}, or as a copper chelate complex. Under reducing conditions, Cu^{2+} may be converted to Cu^{+}, according to the following reaction:

$$Cu^{2+} \leftrightarrow Cu^{+}$$

Mengel and Kirkby (2001) refer that the above transformation is similar to that of Fe, although Cu^{+} is much less stable than the corresponding Fe^{2+} [3]. *Arabidopsis thaliana* possesses a conserved family of CTR-like high affinity Cu transport proteins, denoted as COPT1-5. COPT1 family member participates in plant Cu acquisition [23].

Copper uptake may be influenced by the availability of other nutrients. Antagonism and interactions between Cu and other divalent metals and cations, such as Zn and Fe, are common; according to Graham (1981), Cu-Zn interactions are commonly observed because these metals are absorbed by the same mechanism and therefore, each may competitively inhibit the absorption of the other [24]. A similar interaction is that between Fe and Cu: high Fe availability may reduce Cu absorption [4]. According to Kabata and Pendias (2001), Cu-Cd, Cu-Cr as well as Cu-Mn interactions may occur in the external root media, as well as within plant tissues and may be either synergisitc or antagonistic (especially those of Cu-Cd and Cu-Mn) [4]. Apart from the interactions between Cu with metals having the same valency, other nutrients that may influence Cu uptake are Mo, N and P. It is referred that in some plant species the deficiency of Cu is aggravated by the application of Mo to the soil [25]. Generally, Cu-Mo interactions are closely related to N metabolism; Cu interferes with the role of Mo in the enzymatic reduction of NO_3. The mutual antagonism existing between these nutrients is highly dependent on plant species and the kind of N nutrition [4].

The influence of excess N application on the reduction of Cu concentration of plants is not owed to an interaction between these nutrients, but is the result of increased plant growth due to high N supply (dilution effect). On the other hand, high P levels may decrease Cu availability because insoluble phosphate substances of Cu are formed [16]. In that case, Cu uptake is significantly reduced. Finally, the negative influence effect of Ca and Mg on Cu availability and uptake may be ascribed either to an antagonism in the root surface for the same absorption sites or to the reduction of Cu solubility and availability, through the indirect increase of soil pH.

It has recently been found that an important strategy to maintain Cu homeostasis in *Arabidopsis thaliana* plants consists of regulating uptake and mobilization *via* the conserved family of CTR/COPT Cu transport proteins. More specifically, COPT1 protein mediates root Cu acquisition, whereas COPT5 protein functions in Cu mobilization from intracellular storage organelles. The function of these transporters becomes critical when environmental Cu bioavailability diminishes. Finally, apart from the COPT1 and COPT5 protein functions in Cu acquisition and mobilization, respectively, a very important role has been also found for COPT6 protein in Arabidopsis Cu distribution [26]. In another study, that of Mikkelsen *et al.* (2012), it was found that in barley plants HvHMA1, which is a broad-specificity exporter of metals from chloroplasts and serves as a scavenging mechanism for mobilizing plastid Cu when cells become deficient, might be involved in mobilizing Cu from the aleurone cells during grain filling and germination [27].

Physiological and Biochemical Functions

Copper activates several enzymes having vital functions in plant metabolism, while it also plays a significant role in several physiological processes (photosynthesis, respiration, carbohydrate distribution and metabolism, N reduction and fixation, protein metabolism, superoxide scavenging, cell wall remodelling, ethylene perception, response to pathogens *etc.*) [28, 29]. More specifically, Cu is an important redox agent in biological systems [26]. Being an essential component of many of the electron carriers, Cu is involved in reductive and oxidative electron transport pathways in chloroplasts and mitochondria, respectively [29]. It has been also found that Cu influences water permeability of xylem vessels and thus controls water relationships. Copper controls the production of DNA and RNA and its deficiency greatly inhibits the reproduction of plants (reduced seed production, pollen sterility) [4]. It is also referred that Cu intensifies color and flavor of fruits [30]. Finally, Cu participates in defence against pathogens by involving itself in the cell wall lignification and ascorbic acid metabolism [31].

Cu Deficiency Symptoms and Critical Concentrations

Some of the most characteristic symptoms of Cu deficiency include the formation of narrow and twisted leaves, as well as leaf curling, while their petioles bend downward (Fig. **1**). Recently matured leaves show netted, green veining with areas bleaching to a whitish gray. Some leaves may develop sunken necrotic spots and have a tendency to bend downward [32]. Deficiencies show up first on youngest leaves, young tips, buds and shoots. Copper deficiency causes irregular growth and pale green leaves that wither at leaf margins [30] (Figs. **2, 3**). In cereal crops Cu deficiency shows first in the leaf tips at tillering, although in severe cases it may appear even earlier [3]. When *Citrus* trees are deficient in Cu fruits may show a scarring or rouphened surface (Fig. **4**). A typical Cu deficiency symptom in leaves of *Citrus* trees is shown in Fig. (**5**). The inhibition of lignin synthesis is another disturbance of plant metabolism due to the lack of Cu; reduced lignin synthesis may negatively influence the resistance of plants to diseases from fungi and insects' attack. According to Kabata and Pendias (2001), the resistance of plants to fungal diseases is likely to be related to an adequate Cu supply [4]. Furthermore, low Cu levels resulted in a decrease of the leaf chlorophyll content, as well as in the concentration of reducing sugars in *Hordeum vulgare* L. plants [33].

The critical concentration for Cu deficiency was found to be in the range 4-5 mg/kg d.w. in the 7th to the 9th youngest leaf blades [34]. The deficiency levels of Cu in plants show large genetic differences; however, generally Cu levels less than 2 p.p.m. are considered to be inadequate for plant growth [3]. Jones (1972) refers that the concentration of 4 p.p.m. is the critical limit for Cu deficiency in sensitive plant species [35].

Fig. (1). Copper deficiency symptoms in tomato plants (from Epstein and Bloom, 2005) [36].

Karyotis *et al.* (2011), who studied the variability of nutrients in annual biomass in mountainous and hilly areas of Central Greece, found that Cu concentrations varied within the range from 4.5 to 11 mg/kg d.w., with mean values around 7.40 mg/kg d.w. In perennial pasture crops Cu concentrations varied from 5 to 8 mg/kg d.w. (mean value 6.22 mg/kg d.w.) All these concentrations, according to the same authors, were greater than the critical value of Cu starvation [10].

Under conditions of Cu deficiency in soils, forage species may have low Cu concentrations in their tissues, thus domestic animals feeding on those species may suffer from Cu deficiency. For that purpose, pasture soils should be amended with Cu containing fertilizers and the ruminants should be continuously supplemented with specifically tailored mineral mixture containing enough Cu [37]. Copper deficiency in pasture species, grown in low Cu content soils, as well as in sheeps' blood, at Nisici Plateau of Bosnia-Herzegovina, has been also referred by Manojlovic and Singh (2012) [38].

Fig. (2). Copper deficiency symptoms (holes formed by the expansion of deformed tissue) (**from O' Sullivan,** (http://keys.lucidcentral.org/keys/sweetpotato/key/Sweetpotato%20Diagnotes/Media/Html/TheProblems/MineralDeficiencies/CopperDeficiency/Copper%20%20deficiency.htm) [34].

Detection of Cu Starvation

There are 4 basic methods to diagnose Cu deficiency: i) macroscopic observation of Cu starvation symptoms, ii) soil analysis; however, this is not a reliable tool to diagnose Cu deficiency, since sufficient Cu levels in soil do not guarantee adequate Cu uptake and Cu sufficiency for plants, iii) tissue analysis, iv) biochemical assays. Soil extraction using EDTA with ammonium bicarbonate has been recommended for evaluation of Cu status, particularly in alkaline and calcareous soils. Using this test, a critical concentration of 0.3-0.4 mg/kg Cu has

Fig. (3). Copper deficiency symptoms (from http://tcbanana.blogspot.gr/2011/07/copper-cu-deficiency-symptoms.html) [39].

Fig. (4). Copper deficiency symptoms in *Citrus* fruits (from Horticulture 422: *Citrus* and sub-tropical crops. Mineral nutrition disorders: http://aggie-horticulture.tamu.edu/syllabi/422/pics/nutr/cu.htm) [40].

Fig. (5). A leaf of pineapple sweet orange showing classic mottle typical of copper deficiency (from Stein *et al.*, 2007) [41].

been determined for wheat [34]. Concerning biochemical assays as diagnostic tool for the detection of Cu starvation, it has been found that the activities of polyphenol oxidase, ribonuclease and peroxidase were increased under limited Cu availability [33].

It has been found that *Arabidopsis* antioxidant Protein 1 (ATX1) plays an essential role in Cu homeostasis, conferring tolerance to Cu deficiency. Overexpression of ATX1 resulted in hypersensitivity to severe Cu deficiency [42]. According to Reuter *et al.* (1966), a quick and inexpensive test for the diagnosis of Cu starvation may be the painting of the surface of a young leaf with a dilute solution containing Cu (*e.g.*, 0.25% $CuSO_4.5H_2O$+0.25% $Ca(OH)_2$). If Cu deficiency is the cause of the observed symptoms, the treated leaf will show reduced symptoms and increased expansion of the lamina over the following week. According to the same authors, it is important to clearly label the leaf so that it can be identified on later inspection. The result is referred to be most obviously seen if only one half of the leaf is painted, so that it can be compared with the untreated half [43].

Mechanisms of Tolerance Adopted by Plants to Overcome Cu Starvation

Under conditions of Cu deficiency plants usually enhance remobilization processes in order to overcome the scarcity and survive. It was recently found that COPT5 transport protein in *Arabidopsis thaliana* plays an important role in plant response to Cu scarcity, probably by remobilizing Cu from prevacuolar vesicles, which could act as internal stores to provide the metal cofactor to key Cu-dependent processes, such as photosynthesis [23]. In addition, the antioxidant protein ATX1 was found to play an essential role in Cu homeostasis, in conferring tolerance to Cu deficiency [44].

Furthermore, under Cu deficiency stress (as in any other kind of stress) the production of reactive oxygen species (ROS) is increased and plants are obliged to enhance their antioxidant defence mechanisms (enzymatic and non-enzymatic antioxidant response) in order to survive. Enzymatic and non-enzymatic antioxidant response can alleviate oxidative damage by scavenging ROS. Non-enzymatic antioxidants include low molecular weight compounds, such as glutathione, phenolics and ascorbate, while enzymatic defence includes the activity of antioxidant enzymes, such as superoxide dismutase (SOD), glutathione reductase (GR), catalase, glutathione peroxidase (GPX), ascorbate peroxidase (APX) *etc.* It has been found that under Cu starvation the activities of the enzymes SOD, catalase, GPX, APX and GR increased in *Morus alba* L. plants [29].

Cu Fertilization in Soil and Foliar Application

In order to enhance Cu availability in Cu deficient soils, both chemical fertilizers, as well as organic amendments (such as different kind of manures) can be applied; according to Richards *et al.* (2011), long-term application of biosolids and beef manure significantly increased DTPA extractable Cu concentrations in soils [20]. In addition, it was found that poultry manure applications had a positive effect on Cu concentrations in *Lycopersicon esculentum* L. plants [45].

It is referred that $CuSO_45H_2O$ can be used for foliar application in tomato plants, grown in aquaponic and hydroponic systems [46]. However, it was the unique substance from the micronutrients applied foliarly that did not increase plant fruit number and yield. According to Karamanos *et al.* (2004), foliar applications of Cu were not always as effective as the incorporation of at least 4 kg Cu/ha in the form of $CuSO_45H_2O$ [47]. Apart from Cu sulphate (25% Cu), other sources that could be used as Cu fertilizers are: cupric oxychloride (50% Cu), cuprous oxide (32.8% Cu) and chelated Cu (5% Cu) [48].

Finally, soaking seeds in Cu containing solutions may be a good solution in preventing Cu deficiency; according to Malhi (2009), by soaking wheat seed in Cu EDTA solution at a very low rate, Cu deficiency in wheat may be prevented and seed yield may be increased [49].

REFERENCES

[1]   Kumaragamage D, Indraratne SP. Systematic approach to diagnosing fertility problems in soils of Sri Lanka. Com Soil Sci Plant Anal 2011; 42: 2699-2715.
[2]   Shah Z, Shah MZ, Tariq M, Rahman H, Bakht J, Amanullah M, Shahi M. Survey of *Citrus* orchards for micronutrients deficiency in Swat valley of north western Pakistan. Pakistan J Bot 2012; 44: 705-710.
[3]   Mengel K, Kirkby E. Zinc. In: Mengel K, Kirkby E, Kosegarten H, Appel T (Eds) Principles of Plant Nutrition, 5th ed. Kluwer Academic Publishers, Dordrecht, The Netherlands 2001.
[4]   Kabata A, Pendias H. Trace elements in soils and plants. 3rd ed. 2001; CRC Press, USA 2001.
[5]   Ozyazici MA, Ozyazici G, Dengiz O. Determination of micronutrients in tea plantations in the Eastern Black sea region, Turkey. Afr J Agric Res 2011; 6: 5174-5180.
[6]   Sequeira CH, Baros NF, Lima-Neves JC, Ferreira-Novais R, Silva IR, Alley M. Micronutrient soil test levels and Eucalyptus foliar contents. Com Soil Sci Plant Anal 2011; 42: 475-488.
[7]   Lindsay WL, Norvell WA. Development of a DTPA soil test for zinc, iron, manganese and copper. Soil Sci Soc Am J 1978; 42: 6421-6428.
[8]   Irmak S. Copper correlation of irrigation water, soils and plants in the Cucurova region of Turkey. Int J Soil Sci 2010; 5: 185-195.
[9]   Hitsuda K, Toriyama K, Subarrao GV, Ito O. Percent relative cumulative frequency approach to determine micronutrient deficiencies in soybean. Soil Sci Soc Am J 2010; 74: 2196-2210.
[10]   Karyotis T, Toulios M, Alexiou J, *et al.* Soils and native vegetation in a hilly and mountainous area in Central Greece. Com Soil Sci Plant Anal 2011; 42: 1249-1258.

[11] Somasundaram J, Meena HR, Singh RK, Prasad SN, Parandiyal AK. Diagnosis of micronutrient imbalance in lime crop in semi-arid region of Rajasthan, India. Com Soil Sci Plant Anal 2011; 42: 858-869.

[12] Turan HS, Aydoglu E, Pekcan T, Colakoglu H. Microelement status and soil and plant relationships of olive groves in west Anatolia region of Turkey. Com Soil Sci Plant Anal 2013; 44: 80-88.

[13] Edwards AC, Coull M, Sinclair AH, Walker RL, Watson CA. Elemental status (Cu, Mo, Co, B, S and Zn) of Scottish agricultural soils compared with a soil-based risk assessment. Soil Use Manag 2012; 28: 167-176.

[14] Sinclair AH, Edwards AC. Micronutrient deficiency problems in agricultural crops in Europe (chapter 9). In: Alloway BJ (Ed), Micronutrient deficiencies in global crop production. Springer, Dordrecht 2008; pp. 225-246.

[15] Vogel JG, Jokela EJ. Micronutrient limitations in two managed southern pine stands planted on Florida Spodosols. Soil Sci Soc Am J 2011; 75: 1117-1124.

[16] Alifragis D. Soil: Genesis, Properties and Classification. Volume I. Aivazi Publications 2008; Thessaloniki, Greece. (In Greek).

[17] Pandian SS, Robin S, Vinod KK, *et al*. Influence of intrinsic soil factors on genotype-by-environment interactions governing micronutrient content of milled rice grains. Aus J Crop Sci 2011; 5: 1737-1744.

[18] Sharma BD, Kumar R, Singh B, Sethi M. Micronutrients distribution in salt-affected soils of the Punjab in relation to soil properties. Arch Agron Soil Sci 2009; 55: 367-377.

[19] Xie W, Zhou J, Wang H, Liu Q, Xia J, Lv X. Cu and Pb accumulation in maize (*Zea mays* L.) and soybean (*Glycine max* L.) as affected by N, P and K application. Afr J Agr Res 2011; 6: 1469-1476.

[20] Richards JR, Zhang H, Schroder JL, Hattey JA, Raun WR, Payton ME. Micronutrient availability as affected by the long-term application of phosphorus fertilizer and organic amendments. Soil Sci Soc Am J 2011; 75: 927-939.

[21] Khan F, Iqbal A, Naveedullah A, Khattak MK, Zhou WJ. Physicochemical properties and fertility status of water-eroded soils of Sharkul area of district Mansehra, Pakistan. Soil Environ 2011; 30: 137-145.

[22] Yuan-Rong L, Li-Juan L, Qi-Wen W, Guo-Zhen D. Copper deficiency in Guizhou semi-fine wool sheep on pasture in South West China karst mountain area. Afr J Biotech 2011; 10: 17043-17048.

[23] Garcia-Molina A, Andres-Colas N, Perea-Garcia A, Del Valle-Tascon S, Penarrubia L, Puig S. The intracellular *Arabidopsis* COPT5 transport protein is required for photosynthetic electron transport under severe Cu deficiency. Plant J 2011; 65: 848-860.

[24] Graham RD. Absorption of copper by plant roots. In: Loneragan JF, Robson AD, Graham RD, Eds 'Copper in Soils and Plants', Academic Press, New York, 1981; p. 141.

[25] Olsen SR. Micronutrient interactions. In: Giordano PM, Lindsay WL, Eds 'Micronutrients in Agriculture', Soil Sci Soc Am, Madison, WI, 1972; p. 243.

[26] Garcia-Molina A, Andres-Colas N, Perea-Garcia A, *et al*. The *Arabidopsis* COPT6 transport protein functions in Cu distribution under Cu-deficient conditions. Plant Cell Physiol 2013; 54: 1378-1390.

[27] Mikkelsen MD, Pedas P, Schiller M, *et al*. Barley HvHMA1 is a heavy-metal pump involved in mobilizing organellar Zn and Cu and plays a role in metal loading into grains. PloS ONE 2012; 7: Article number 49027.

[28]   Yuan M, Chu Z, Li X, Xu C, Wang S. The bacterial pathogen *Xanthomonas oryzae* overcomes rice defenses by regulating host Cu redistribution. Plant Cell 2010; 22: 3164-3176.

[29]   Tewari RK, Kumar P, Sharma PN. Antioxidant responses to enhanced generation of superoxide anion radical and hydrogen peroxide in the copper-stressed mulberry plants. Planta 2006; 223: 1145-1153.

[30]   Advanced Nutrients: Hydroponic Articles. Everything you need to know about fixing nutrient problems. Internet: http://www.advancednutrients.com/hydroponics/articles/hydro ponics-gardening/know-about-fixing-nutrient-problems.php)

[31]   Shingles R, Wimmers LE, McCarty RE. Copper transport across pea thylakoid membranes. Plant Physiol 2004; 135: 145-151.

[32]   Taiz L, Zeiger E. Plant Physiology online. A companion to plant physiology, Fifth Edition, Sinauer Associates, Inc. 2010. Internet: http://5e.plantphys.net/article.php?ch=t&id=289

[33]   Chatterjee C, Nautiyal N. Variation in calcium levels leads to changes in the copper metabolism in barley. Soil Sci Plant Nutr 2001; 47: 9-16.

[34]   O' Sullivan J. Copper deficiency. Internet: http://keys.lucidcentral.org/keys/sweetpotato/ke y/Sweetpotato%20Diagnotes/Media/Html/TheProblems/MineralDeficiencies/CopperDefici ency/Copper%20%20deficiency.htm).

[35]   Jones JB. Plant tissue analysis for micronutrients. In: Mordvedt JJ, Giordano PM, Lindsay WL, Eds 'Micronutrients in Agriculture', WI, USA 1972.

[36]   Epstein E, Bloom AJ. Mineral nutrition of Plants: Principles and Perspectives. Sinauer Associates, Inc publishers. Sunderland Massachusetts, 2nd ed. 2005; pp. 225-227.

[37]   Ahmad K, Khan ZI, Shaheen M, Seidavi A. Dynamics of magnesium, copper and zinc from soil to forages, grown in semiarid area in Sargodha, Pakistan. Leg Res 2012; 35: 294-302.

[38]   Manojlovic M, Singh BR. Trace elements in soils and food chains of the Balkan region. Acta Agric Scand B: Soil Plant Sci 2012; 62: 673-695.

[39]   Tissue culture banana cultivation technology. Internet : http://tcbanana.blogspot.gr/2011/0 7/copper-cu-deficiency-symptoms.html

[40]   Horticulture 422: *Citrus* and subtropical crops. Mineral nutrition disorders. Internet: http://aggie-horticulture.tamu.edu/syllabi/422/pics/nutr/cu.htm

[41]   Stein B, Figueroa J, Foguet L, Figueroa A, Escobar C. The *Citrus* sanitation center of the Obispo Colombres Experimental station, Tucumaan, Argentina. Proceedings of the 17th Conference IOCV, IOCV, Riverside, 2008.

[42]   Shin LJ, Yeh KC. Overexpression of Arabidopsis ATX1 retards plant growth under severe Cu deficiency. Plant Sign Behav 2012; 7: 1082-1083.

[43]   Reuther W, Labanauskas CL. Copper. In: Chapman HD, Ed 'Diagnostic criteria for plants and soils' Department of Soils and Plant Nutrition, University of California *Citrus* Research Centre and Agricultural Experiment Station, Riverside, California 1966; pp 157-179.

[44]   Shin LJ, Lo JC, Yeh KC. Copper chaperone antioxidant Protein1 is essential for copper homeostasis. Plant Physiol 2012; 159: 1099-1110.

[45]   Demir K, Sahin O, Kadioglu YK, Pilbeam DJ, Gunes A. Essential and non-essential element composition of tomato plants fertilized with poultry manure. Sci Hortic 2010; 127: 16-22.

[46]   Roosta HR, Hamidpour M. Effects of foliar application of some macro- and micro-nutrients on tomato plants in aquaponic and hydroponic systems. Sci Hortic 2011; 129: 396-402.

[47]   Karamanos RE, Pomarenski Q, Goh TB, Flore NA. The effect of Cu foliar application on grain yield and quantity of wheat. Can J Plant Sci 2004; 84: 47-56.

[48]   Merlin TPA, Lima GPP, Leonel S, Vianello S. Peroxidase activity and total phenol content in *Citrus* cuttings treated with different copper sources. South Afr J Bot 2012; 83: 159-164.

[49]   Malhi SS. Effectiveness of seed-soaked Cu, autumn- *versus* spring- applied Cu, and Cu-treated P fertilizer on seed yield of wheat and residual nitrate-N for a Cu-deficient soil. Can J Plant Sci 2009; 89: 1017-1030.

CHAPTER 7

Boron Deficiency

Abstract: Boron is a very important micronutrient, playing a crucial role in many physiological and biochemical functions, as well as in plant metabolism. More specifically, B is involved in cell elongation and division, cell wall biosynthesis and structure, N, carbohydrate and IAA metabolism, photosynthesis, as well as in membrane integrity, seed production, sugar metabolism, regulation of lignin biosynthesis *etc.* Flowering and fruit setting are two of the mostly known functions that are negatively influenced by B deficiency.

Boron deficiency is most probably found in calcareous and alkaline soils, as well as in soils formed on parent materials inherently low in B, such as sandstones. Boron is absorbed from soil solution mainly as undissociated boric acid. Boron uptake is not yet clear as to the extent to which this process is either passive or active; however, the metabolically controlled process seems to be relatively minor. The most possible explanation is that when B supply is high, B uptake by roots is believed to occur by passive diffusion. In contrast, under low B supply, a significant portion of B may be taken up *via* active pathways. Under low B conditions (leaf concentrations of less than 10 mg/kg dry weight in young sampled expanding leaves), flower sterility and floral abnormalities are within the first symptoms of B starvation. In addition, the formation of incomplete or damaged embryos and malformed fruits are also within the most common and important symptoms of B starvation for plants. Other symptoms of B deficiency include rapid cessation of root elongation, inhibited growth and thickening of young leaves, loss of apical dominance in trees *etc.*

Under B deficiency conditions some tolerance and adaptation mechanisms, such as enhanced translocation from root system to leaves and lower shoot/root dry weight ratio in resistant genotypes, usually take place. In order to alleviate B deficiency, the most commonly used B fertilizer is borax; nevertheless, its' basic drawback is that it leaches easily from sandy soils. Other fertilizers that can be used, either as soil, or foliar application, are those of boric acid and solubor 20-21% (leafy sprays are particularly preferred when soil is potentially capable of fixing high amounts of B).

All these topics concerning B availability in soils and the factors influencing it, the uptake and transport of B, the roles of B in plant metabolism and growth, the critical concentrations of B starvation in plant tissues, the symptoms of B deficiency and the methods of its' detection, as well as the tolerance mechanisms adopted by plants in order to face B deprivation and the B-fertilizers used to alleviate B stress in crops are fully analyzed and discussed in this chapter.

Keywords: B availability, B deficiency, B fertilizers, B mobility, B uptake, borax, boric acid, flower sterility, fruit setting.

Theocharis Chatzistathis

INTRODUCTION

Boron deficiency occurs over a much wider range of soils and crops in comparison to any other micronutrient deficiency; it has been reported in 80 countries and 132 crops around the world [1]. Boron is a highly important micronutrient; it plays a crucial role in flowering and fruit setting. Particularly, B deficiency is responsible for creating male flower sterility and floral abnormalities [2]. It is also involved in several biochemical and physiological processes, such as cell elongation and division, cell wall biosynthesis and structure, N, carbohydrate and IAA metabolism, photosynthesis, as well as in membrane integrity, seed production, sugar metabolism, regulation of lignin biosynthesis and xylem differentiation [3-5]. According to Bolanos *et al.* (2004), B is implicated in three main processes: keeping cell wall structure, maintaining membrane function and supporting metabolic activities [6]. It has been also reported that B deficiency induces stress-responsive genes [4]. Under B deficient conditions, deformation of younger leaves, death of the apical meristem, as well as swelling of the middle lamellae and absence of starch granules may be observed [7]. In addition, rapid cessation of root elongation, inhibited growth of young leaves and reduced fertility are within the most important deficiency symptoms [3].

Boron availability is decreased under alkaline conditions due to its high adsorption in soils [8]. In that case, calcium carbonate acts as an important B adsorbing surface in calcareous soils. Other important soil factors contributing to B adsorption are: pH, organic matter, the kind of clay minerals, the content of Fe and Al oxides/hydroxides *etc.* [9-12]. Boron is taken up from soil solution mainly as undissociated boric acid, and thus in a form potentially permeable to plant cells; generally, B uptake and transport is primarily determined by its' concentration in soil solution and the transpiration rates of plants [12, 13]. Boron is a relatively immobile nutrient and analysis of mature leaves largely reflects B accumulation by transpiration and not the current B supply. However, as a general rule, when B concentrations (in young sampled expanding leaves) are below 10 p.p.m. plants suffer from B deficiency [12]. Under conditions of B deficiency, enhanced translocation from root system to leaves usually takes place and genotypic differences in B retranslocation usually reflects tolerance mechanism to B deprivation [14]. However, most plant species can not retranslocate B efficiently, and therefore require a continuous supply of B throughout their life cycle [15]. For that purpose, the most commonly used B fertilizers are borax and boric acid; however, borax leaches easily from sandy soils. In that case, boric acid may be used either as soil fertilizer or as leafy spray (especially in the cases when soil is potentially capable of fixing high amounts of B) [12].

The purpose of this chapter is to present and discuss all the topics referring to: i) the factors influencing B adsorption and availability in soils, ii) B uptake and translocation by plants, iii) the roles of B in plant physiology and biochemistry, iv) the critical B concentrations and symptoms of starvation, v) the tolerance mechanisms and adjustment strategies adopted by plants in order to overcome B deprivation and, finally vi) the B-fertilizers that can be used by farmers to alleviate B deficiency symptoms.

B in Soils

Boron is the 51st most common element found in the earth's crust and it is found at an average concentration of 8 mg/kg. Boron concentration in rocks ranged between 10-20 mg/kg, while the primary minerals of B are ulexite, borax and natural boric acid [16]. The total B concentration in soils is within the range from 20 to 200 mg/kg d.w., most of which is inaccessible to plants [12]. From the soil types influencing B starvation for crops, Podzols are within the most prone ones, due to their highly leached course texture and to the fact that they are often formed on parent materials inherently low in B, such as sandstones and acid igneous rocks [12]. According to Shorrocks (1997), soil orders with prevalent B deficiency are ultisols, lithic inceptisols, lithic fluvents, alfisols, psamments, oxisols, spodosols and andepts [17]. The less soluble colemanite is released as a result of leaching and weathering, while high pressure produces kernite, which is a less hydrated form of borax [16].

From the B-containing soil minerals, we should distinguish the very insoluble (such as tourmaline, containing 30-40 mg B/kg) and the very soluble hydrated B minerals. Soluble B consists mainly of boric acid ($B(OH)_3$). Apart from the boric acid, another very common form of B in soil solution is $B(OH)_4^-$. Only at pH above 7 there are other anions, such as $H_2BO_3^-$ and $B_4O_7^{2-}$, likely to occur in soil solutions. Soluble forms of B are easily available to plants, which can take up undissociated (non-ionic) boric acid, as well as other B forms present in the ambient solution [13]. Boron is present in soils as undissociated boric acid in a great pH range (from 4 to 8). Thus, B unlike other essential plant nutrients, which are present in ionized form, is mainly present in soils as undissociated boric acid. The fact that boric acid is undissociated in soil solution is the main reason why B can be so easily leached from soils [12].

Soil pH is probably the most important factor influencing B availability; under alkaline soil conditions B availability is significantly decreased. In alkaline soils B adsorption to soil constituents, such as Al and Fe oxides, clay minerals, calcium carbonate and organic matter, is usually enhanced. In general, increasing pH up to about 8-10 favours adsorption [12]. In addition, B deficiency is also favoured in areas with high rainfalls or with temperate climate [18, 19]. In contrast to the above, a weak relationship between organic C and B was found [20].

There are different soil tests that could be used to estimate B available for plants content in soils; DTPA is probabaly the mostly used extractant, as happens with the other micronutrients. DTPA extractable B concentrations within the range from 1.6 to 2.3 are considered as adequate for most crops, according to Richards *et al.* (2011) [21]. Very low B concentrations (in the deficiency range) are those being <0.3 mg/kg, low are considered those varying from 0.3 to 0.5 mg/kg, moderate are the concentrations from 0.5 to 1.0 mg/kg and adequate to high (no risk of deficiency) those exceeding 1.0 mg/kg (until 3.5 mg/kg) [22]. Apart from DTPA, other soil test extractants may be also used for B extractability, such as hot water, HCl and mannitol. According to Rashid *et al.* (1997), between hot water, HCl and mannitol, all the three extractants were found to be almost equally effective in evaluating soil B availability [23].

Factors Affecting B Availability

I. **Soil pH.** Boron adsorption increases as pH increased from 3 to 9 [11, 24]. In contrast, B adsorption decreases with the increase of pH from 10 to 11.5 [25]. According to Mengel and Kirkby (2001), the lower rate of B removal *via* leaching from neutral and alkaline soils is also a consequence of enhanced B adsorption at higher pH values [12].

II. **Soil texture.** Boron deficiency is commonly found in sandy soils, than in clayey ones [26]. On the other hand, with the increase of clay content B adsorption is increased, thus its availability is reduced.

III. **CaCO₃ content.** As $CaCO_3$ content increases, soil pH is also increased and B availability is considerably decreased. Apart from this indirect influence of $CaCO_3$ on B availability through the increase of pH, the direct influence consists of an increased adsorption of B on the surface soil $CaCO_3$ particles [10].

IV. **Organic matter.** As the organic C increases B adsorption is enhanced, thus leaching is reduced. According to Lehto *et al.* (2010), B output from forest ecosystems with potential for leaching is controlled by adsorption in soil, which is poorly understood, particularly in soils with abundant organic matter [27]. According to Kabata and Pendias (2001), soil organic matter adsorbs more B than mineral soil constituents [13]. However, B adsorption on both organic soils and composted organic matter also increased with increasing pH. Finally, on a weight basis the sorption capacity for B in composted organic matter is about 4 times greater than for clay [12].

V. **Salinity.** High salinity levels in soil may decrease B availability for plants; Wimmer and Goldbach (2012) found that B uptake rates were reduced with increasing salt concentration, but only under high B supply [28].

VI. **The kind of clay minerals.** Between the different clay minerals great differences concerning the ability of B adsorption exist; it has been found that B adsorption per unit weight of clay follows the order kaolinite>montmorillonite>illite [12].

VII. **The content of Fe and Al oxides.** Boron adsorption on oxides of Fe and Al is believed to be an important mechanism governing B solubility in soils [9].

VIII. **Precipitations and soil moisture.** In rainy areas, and especially in sandy soils receiving exaggerate irrigations, B leaching may be a reality. Lehto *et al.* (2010) refer that B deficiencies are widespread in regions with sandy soils, coupled with strong leaching rainfall regimes [27]. The fact that B is mainly present in soils as undissociated boric acid is the reason why B can be so easily leached from soils [12].

IX.   **The addition of lime or gypsum.** Experimental liming was found to greatly reduce B availability in soils; this is one of the reasons, according to Lehto and Malkonen (1994), why liming never became a practical forest management tool in Finland [29]. In addition, heavy applications of $Ca(H_2PO_4)_2$ resulted, according to Kabata and Pendias (2001), in a lower availability of B [13].

Boron Uptake, Transport and Mobility

There is still controversy as to the extent to which B uptake process is either passive or active; the metabolically controlled process seems to be relatively minor. However, in sunflower plants the high uptake efficiency values found under low B concentrations indicated that mechanisms other than mass flow played a role in providing the acquired B to plants [30]. Generally, it seems that at low B supply its' accumulation in the symplast of root cells is considered to depend on two processes, working together: i) energy-dependent process, and ii) a passive-diffusion process along a gradient, maintained by the formation of B complexes within the cell [31]. In contrast, when B supply is high, B uptake by roots is believed to occur by passive diffusion. According to Wimmer and Goldbach (2012), under low B supply, when a significant portion of B can be taken *via* active pathways, transpiration is not the decisive factor for B accumulation. However, transpiration-water driven flow and diffusion are the dominant factors for B accumulation in aerial plant parts under high B supply [28]. Boron is absorbed from soil solution mainly as undissociated boric acid, and thus in a form potentially permeable to plant cells [12]. Boric acid uptake was found to be facilitated by membrane intrinsic proteins in *Cucurbita pepo* roots [32, 33]. It can be predicted that B uptake is primarily determined by the B concentration of soil solution and the transpiration rate of plants; however, B uptake also differs between species, genotypes *etc.* [13, 34]. Chilling temperatures in the root environment restricts B uptake capacity and distribution utilization/efficiency in the shoot. So, for subtropical/tropical species (cucumber, cassava, sunflower) root chilling at 10-17°C decreases B uptake efficiency and utilization in the shoots, while it also increases the shoot: root ratio; chilling-tolerant temperate species require much lower root chill temperatures (2-5°C) to achieve the same responses [35].

In plants adequately supplied with B, as much as 60% of the boron's total content can be present in soluble form [36, 37]. It has been found that B is relatively mobile. In *Eucalyptus* for example (especially in clone 129), its higher mobility could be due to the presence of an organic compound, such as mannitol, able to complex B [14]. Species including garlic, celery, asparagus, cauliflower, carrot, olive trees, beans, peas and coffee contain mannitol, which is capable of forming phloem transportable complexes of B [38].

Roles of Complexes of B in Plant Physiology, Biochemistry and Metabolism

According to Bolanos *et al.* (2004), B is implicated in three main processes: keeping cell wall structure, maintaining membrane function and supporting metabolic activities [6]. More specifically, the roles of B in plants include cell division and elongation, sugar transport, cell wall synthesis, lignification, carbohydrate, ascorbate, phenol, RNA and IAA metabolism, protein synthesis, nitrogen fixation, movement of carbohydrates across cell walls, satiability of pollen germ tube, metabolism of phenolics, as well as amelioration of Al toxicity [12, 13, 39].

The chemical composition and ultrastructure of cell walls are quickly affected by a lack of B. Cell walls become thicker in root apical meristems. The swelling of cell walls under B deprivation and its relationship with the borate-ester cross-linked rhamnogalacturonan II dimer (RG-II) was firstly described by Ishii *et al.* (2001) [40]. According to Da Silva *et al.* (2008), the presence of RG-II has been verified in cell walls of many species, belonging to different families, such as *Brassicaceae, Cucurbitaceae, Leguminosae, Solanaceae etc.* From *Euphorbiaceae* family, *Ricinus communis* was the first species in which verified the existence of a RGII complex [7].

Cessation of root elongation is one of the most rapid responses to B deficiency [12]. This happens due to the limiting of cell enlargement and cell division in the growing zone of root tips [41]. Not only root tip cessation due to B starvation occurs, but also mycorrhiza development is impaired, and both are considered to be the first phenomena affected by B supply shortage [27]. In a field study, the number of ectomycorrhizas per unit soil volume in Norway spruce has been increased after B fertilization. This probably happened due to the increased number of short root tips, which are available for mycorrhizal colonization [42]. Although the enhancement of mycorrhiza formation by increased B availability is puzzling, one possible explanation could be the increased carbohydrate allocation; another reason for the increased mycorrhiza colonization could be that mycorrhizal fungi may require a crucial amount of B for their growth [27].

Boron plays a great role in flower formation, sexual reproduction and fruit setting. In flowers, low B reduces male fertility, primarily by impairing microsporogenesis and pollen tube growth; impaired embryogenesis, resulting in seed abortion, or the formation of incomplete or damaged embryos and malformed fruits are within the most common and important symptoms of B starvation for plants [41].

Boron has been found to be essential for the N_2 fixation symbiosis of *Rhizobium* and *Frankia* [6]. Finally, B mineral nutrition was also found to play a significant role in the induced defense of birches (resistance of seedlings to larvae of the autumnal moth *Epirrita autumnata*) [43].

Critical Concentrations and Symptoms of B Deficiency

Boron is relatively immobile and analysis of mature leaves largely reflects B accumulation by transpiration and not the current B supply. This can be only achieved by sampling growing tissues. Generally, leaf concentrations of less than 10 mg/kg dry weight are associated with deficiency symptoms in young sampled expanding leaves [12]. It has been relatively easy to establish the critical B levels in plant tissues as 5 to 30 p.p.m. For example, the average B concentration in grasses is 5.7 p.p.m. The range of B content in vegetables and fruits varies from 1.3 to 16 p.p.m. [13]. In Table **1** are presented the mean B contents of grasses, legumes, vegetables and cereal grains from different countries. According to Da Silva *et al.* (2008), the mean B content in new leaves of castor bean (*Ricinus communis*) plants were found to be 12 mg/kg dry weight in the B absent treatment. The same authors refer that B deficiency symptoms, such as deformity and necrosis of leaf edges (Figs. **1**, **2**), firstly appeared in new leaves at the 40th day after transplanting, and also that these symptoms were probably owed to low or restricted B phloem mobility in this species (*Ricinus communis*) [7].

The symptoms of B deficiency include rapid cessation of root elongation, inhibited growth of young leaves and reduced fertility [3]. The first sign is usually a thickening of young leaves. The leaves and stem near the shoot tip are brittle and break easily when bent [44]. According to Da Silva *et al.* (2008), the major symptom of B deficiency was the thickening of the middle lamellae. That fairly and rigid layer is a structural component, located between two adjacent primary cell walls, composed of pectin. The thickening of the middle lamellae could be possibly explained by the structural role of B in relation to the polysaccharide present in the pectin [7].

Moreover, several physiological processes, such as sugar transport, cell wall synthesis, lignification, carbohydrate, RNA and IAA metabolism are seriously disturbed due to B starvation [15]. Boron deficiency may also result in stunted plant growth. In the study of Lehto *et al.* (2010) it was found that B deficiency caused loss of apical dominance in trees. According to the same authors, if this is repeated for many years it may lead to study, bushy appearance of trees [27]. In addition, seed and grain production are also reduced with low B supply. Da Silva *et al.* (2008) refer that the seed production of castor bean *(Ricinus communis)* plants was strongly affected by B absence [7]. According to Dell and Huang (1997), the early inhibition of root growth, compared to shoot growth, increases the shoot/root ratio; this may enhance the susceptibility of plants to environmental stresses, such as marginally deficient supplies of other nutrients and water deficit [41].

In fruits, fibrous roots of B-deficient sweet-potato plants become short (Fig. **3**), stumpy and highly branched [44]. According to Lehto *et al.* (2010), the impaired development of the primary cell wall in B-deficient trees provoke negative

Table 1. Boron contents of grasses, legumes and cereal grains (p.p.m. of dry weight) from different countries (modified from Kabata and Pendias, 2001) [13].

Country	Grasses		Clovers	
	Range	Mean	Range	Mean
Great Britain	-	26	-	-
Czech Republic	14-30	22	-	-
Finland	3.9-6.3	4.9	-	-
Germany	-	-	20-50	-
Hungary	1.0-7.9	5.8	20-35	33
Japan	1.6-12	4.9	12-35	21
Poland	1.0-15.6	5.6	11.3-16.5	14
New Zealand	1.7-10	5.2	6-120	26
U.S.	<5-20	7.4	10-70	22
Russia	2-10	5	32-50	40
Yugoslavia	-	-	70-97	78

Country	Cereal Grains		
	Kind of Cereal	Range	Mean
Canada	Oats	-	0.7
Great Britain	Barley	-	3.4
	Oats	-	3.3
Poland	Oats	1.9-2.3	2.0
	Rye	1.1-1.6	1.3
	Wheat	0.3-1.5	0.8
Finland	Barley	0.7-1.1	1.0
	Oats	0.9-1.3	1.2
	Wheat	1.1-1.2	1.2
U.S.	Barley	0.9-2.3	1.6
	Oats	1.6-3.8	2.2
	Wheat	0.8-4.3	1.9
Russia	Barley	5-9	6.6
	Oats	5-8	6.8
	Wheat	1-15	6.8

consequences and disorders in the structural development of organs and whole plants. This usually negatively affects wood quality and productivity of forest tree species. According to the same authors, at least part of the value of wood production is lost by the time B deficiency symptoms appear [27].

Great differences exist between plant species concerning crop B requirements and tolerance to B deprivation. Some of the most sensitive to B deficiency species include *Cruciferae, i.e.,* cabbage, turnips, brussels, sprouts, cauliflower and some of the *Chenopodiaceae,* such as sugar beet [12]. From tree crops, apple tree (*Malus domestica*) is known to be sensitive to B deprivation due to its high B requirements [45]. Other crops sensitive to B deficiency include celery, coffee, oil palm, cotton, sunflower, olive tree and pine. Between Graminaceous monocots and dicots, both plant B requirements and B concentrations in tissues are much lower in monocots, than in dicots; it seems that differences in cell wall composition and lower amount of pectic substances in the *Graminaceae* are probably responsible for these differential B requirements and contents [12]. The ratio of B in dicots to B in monocots was found to be 6.5 [46].

Fig. (1). Boron deficiency symptoms in the leaf of an oilseed rape of canola plant (*Brassica napus*) [47].

Fig. (2). Older leaves on B deficient plants develop a yellow border (top and left). New leaves are distorted and appear mottled (right) [48].

Fig. (3). Short, blunt-ended roots from a B-deficient crop (left) and normal spindle-shaped roots (right), developing on a younger crop to which boron was applied (L. Loader) [44].

Diagnostic Tools

There are different methods of diagnosing B deficiency: the recognition from the appearance of starvation symptoms is probably the easiest one; however, it demands a lot of experience, since it may be confused with other nutrient deficiencies/disorders or fungi/insects' attack. For example, many coniferous species are sensitive to loss of apical dominance for a number of other reasons, such as insect and mammal herbivory. Therefore, the visible symptoms of stunted growth do not correspond to low B concentrations to encourage the use of growth disturbance as a key diagnostic tool. Furthermore, after the appearance of deprivation symptoms it is too late to alleviate B deficiency (as in any other nutrient deficiency), since plant metabolism has already been seriously disturbed. In addition to the recognition of B deprivation symptoms in leaves and/or fruits, a soil analysis with a suitable B extractant (like DTPA) in order to determine B available levels in soils, together with a leaf analysis, are always a 'must' which will help the farmer to ensure the detection of B deficiency. It should not be forgotten that B is relatively immobile and B analysis of mature leaves largely reflects its' accumulation by transpiration and not the current B supply. This problem can be only solved by also sampling growing tissues [12, 44].

Interactions with Other Nutrients

Boron application decreased tissue N, Ca and Mg, but increased tissue P, K, Fe, Mn, Zn and Cu concentrations in *Capsicum annum* L. and *Cucumis sativus* L. plants [49]. In similar results concluded the research of Esringu *et al.* (2011) for strawberry plants [50]. The B-Ca interrelationship is most often reported; lime-induced B deficiency has been frequently observed in acid soils. However, B starvation is not always a case in limed acid soils. Indeed, it has been found that despite the fact that liming decreased soil B levels in an acidic soil, at the same time it did not affect B concentration in *Malus domestica* trees and accelerated the uptake of added B, indicating a possibility for increased soil-to-plant mobility of B. The increased uptake of B after liming should be ascribed to the improved growth conditions that were established, which in turn improved B absorption by plants [51]. In addition, B is reported to have beneficial effects on Al toxicity;

however, the ameliorative effect of B is related rather to root activity under Al stress and not directly to Al toxicity [13].

Tolerance to B Deficiency Mechanisms

Great differences among plant species concerning sensitivity to B deprivation occur; for example, coffee tree is one of the most sensitive species to B deficiency [52]. Sunflower is also considered a susceptible crop, in which B constraint produces a wide variety of symptoms [30]. According to Liu *et al.* (2013), differences in B uptake, cellular B allocation and pectin content can explain genotypic differences in B efficiency between B-efficient and B-inefficient *Citrus* rootstocks. A 28% reduction of shoot dry mass in trifoliate orange plants was found after 35 days under B deprivation, but shoot dry mass of citrange was not significantly affected [34]. Genotypic differences in tolerance to B starvation does not mean that differences also exist in B content among genotypes; between a sensitive to B deficiency sunflower genotype and a tolerant one, the total B content did not significantly differ between them. This finding probably suggested that the sensitive genotype exhibited a greater B requirement for root cell wall than the resistant [30].

Under B deficiency conditions enhanced translocation from root system to leaves usually takes place. Less sensitive to B-deficiency clones of *Eucalyptus* owe their tolerance to B deficiency stress to a higher ability to retranslocate B from root system to shoots, than the more sensitive ones [14]. However, most plant species can not retranslocate B efficiently, and therefore require a continuous supply of B throughout their life cycle [15]. According to Lehto *et al.* (2010), B remobilization within trees may be a key factor in the occurrence of forests in areas with very low B availability. However, the ability to remobilize B varies considerably between plant species [27].

Another B deprivation mechanism is that the shoot/root dry weight is usually lower in resistant genotypes; according to Fuertes *et al.* (2010) the shoot/root dry weight ratio was lower in a resistant to B deficiency sunflower genotype, than in a sensitive one. This means that this genotype would have a substantially larger root volume, capable of supporting the B demand of its shoots. Such an adjustment in shoot/root ratio could be part of an effective strategy, enabling sunflower plants to take up more B in order to satisfy the B demands of shoots when B supply is restrictive under field conditions [30]. Root active uptake and remobilization of B have been accepted as mechanisms contributing to nutrient efficiency under low B supply [53]. It has been found that CiNIP5 gene is responsible for B uptake into the root under B-deficient conditions; furthermore, the expression of this gene is related to tolerance to B deprivation between the roots of *Carrizo citrange* and *Fragrant citrus*, in which the former showed more tolerance than the latter to low B conditions in field practice [54].

Boron Fertilizers

Differences concerning solubility exist between B-containing fertilizers. In a study conducted between 4 boron fertilizer materials (Granubor, Hydroboracite, Ulexite and Colemanite) used for fertilization of *Medicago sativa* L., it was found that plants grown with Granubor and Ulexite had higher B concentrations, than those grown with Hydroboracite, showing higher solubility than Colemanite [55]. According to Mengel and Kirkby (2001), the natural ore Colemanite can be used similarly to avoid leaching losses, provided that the particle size is small enough [12]. Between powder and granular colemanite it has been found that the first was more efficient than the second in supplying B to rice due to its finer particle size, making it more water soluble [1]. However, the most commonly used B fertilizer is borax, but it leaches easily from sandy soils. Differences between B-fertilizers exist not only concerning their solubility, but also concerning their B concentration; for example, borax contains 11% B, boric acid 17% and solubor 20-21% [12]. According to Mengel and Kirkby (2001) boric acid is also frequently applied as leafy spray, particularly when the soil is potentially capable of fixing high amounts of B. A common problem of B leaf application, not encountered with other micronutrients, is the limited range between conditions of deficiency and toxicity [12]. This means that plants in B-deficient soils may exhibit toxicity symptoms if they are over-fertilized with B [51].

The addition of manures and other organic amendments is another way of increasing B available levels in soils. Beef manure, for example, was found to contain B from 77 to 201 mg/kg, with a mean concentration of 129 mg/kg; it was found that the long-term application of beef manure leaded to a significant linear increase in DTPA extractable B [21].

REFERENCES

[1]   Saleem M, Yusop MK, Ishak F, Samsuri AW, Hafeez B. Boron fertilizers borax and colemanite application on rice and their residual effect on the following crop cycle. Soil Sci Plant Nutr 2011; 57: 403-410.

[2]   Sharma CP. 2006. Plant micronutrients. Enfield, USA: Science Publishers, 2006.

[3]   Marschner H. Mineral Nutrition of Higher Plants. 2nd edition. Academic Press, London 1995.

[4]   Gonzalez-Fontes A, Rexach J, Navarro-Gochicoa MT, *et al.* Is boron involved solely in structural roles in vascular plants? Plant Signal Behav 2008; 3: 24-26.

[5]   Gupta UC, Srivastava PC, Gupta SC. Role of micronutrients: Boron and molybdenum in crops and in human health and nutrition. Curr Nutr Food Sci 2011; 7: 126-136.

[6]   Bolanos L, Lukaszewski K, Bonilla I, Blevins D. Why boron? Plant Physiol Biochem 2004; 42: 907-912.

[7]   Da Silva DH, Rossi ML, Boaretto AE, De Lima-Nogueira N, Muraoka T. Boron affects the growth and ultrastructure of castor bean plants. Sci Agric 2008; 65: 659-664.

[8]   Bhakuni G, Khurana N, Chatterjee C. Impact of boron deficiency on changes in biochemical attributes, yield and seed reserves in chickpea. Com Soil Sci Plant Anal 2010; 41: 199-206.

[9]   Lindsay WL. Inorganic phase equilibria of micronutrients in soils. In: Mortvedt JJ, Giordano PM, Lindsay WL (Eds) Micronutrients in Agriculture. Soil Sci Soc Am, Madison, WI 1972; p. 41.

[10] Goldberg S, Forster HS. Boron sorption on calcareous soils and reference calcites. Soil Sci 1991; 152: 304-310.

[11] Lehto L. Boron retention in limed forest. For Ecol Manag 1995; 78: 11-20.

[12] Mengel K, Kirkby E. Zinc. In: Mengel K, Kirkby E, Kosegarten H, Appel T (Eds) Principles of Plant Nutrition, 5th Edition 2001; Kluwer Academic Publishers, Dordrecht, The Netherlands, 2001.

[13] Kabata-Pendias A, Pendias H. Trace Elements in Soils and Plants. 3rd ed. CRC Press, USA, 2001.

[14] Jose JFBS, Da Silva IR, De Barros NF, *et al.* Boron mobility in *Eucalyptus* clones. Rev Bras Cien Solo 2009; 33: 1733-1744.

[15] Koshiba T, Kobayashi M, Matoh T. Boron nutrition of tobacco BY-2 cells. Oxidative damage is the major cause of cell death induced by boron deprivation. Plant Cell Physiol 2009; 50: 26-36.

[16] Shaaban MM. Role of boron in plant nutrition and human health. Am J Plant Physiol 2010; 5: 224-240.

[17] Shorrocks VM. The occurrence and correction of boron deficiency. Plant Soil 1997; 193: 121-148.

[18] Stevens G, Dunn D. Fly-ash as a liming material for cotton. J Environ Qual 2004; 33: 343-348.

[19] Fageria NK, Baligar VC, Zobel RW. Yield, nutrient uptake and soil chemical properties as influenced by liming and boron application in common-bean in a no-tillage system. Com Soil Sci Plant Anal 2007; 38: 1637-1653.

[20] Karyotis T, Haroulis A, Vavoulidou E, Papadopoulos P. Soil properties and distribution of heavy metals and boron within three Greek histosols. Suo 2000; 51: 95-104.

[21] Richards JR, Zhang H, Schroder JL, Hattey JA, Raun WR, Payton ME. Micronutrient availability as affected by the long-term application of phosphorus fertilizer and organic amendments. Soil Sci Soc Am J 2011; 75: 927-939.

[22] Edwards AC, Coull M, Sinclair AH, Walker RL, Watson CA. Elemental status (Cu, Mo, Co, B, S and Zn) of Scottish agricultural soils compared with a soil-based risk assessment. Soil Use Manag 2012; 28: 167-176.

[23] Rashid A, Rafique E, Bughio N. Micronutrient deficiencies in rainfed calcareous soils of Pakistan. III. Boron nutrition of sorghum. Com Soil Sci Plant Anal 1997; 28: 441-454.

[24] Barrow N.J. Testing a mechanistic model. X. The effect of pH and electrolyte concentration on borate sorption by a soil. J Soil Sci 1989; 40: 427-435.

[25] Goldberg S, Glaubig RA. Boron adsorption on California soils. Soil Sci Soc Am J 1986; 50: 1173-1776.

[26] Shaaban MM, Abdalla FE, Abou El-Nour EAA, El-Saady AM. Boron/Nitrogen interaction effect on growth and yield of faba bean plants grown under sandy soil conditions. Int J Agric Res 2006; 1: 322-330.

[27] Lehto T, Ruuhola T, Dell B. Boron in forest trees and forest ecosystems. For Ecol Manag 2010; 260: 2053-2069.

[28] Wimmer MA, Goldbach HE. Boron and salt interactions in wheat are affected by boron supply. J Plant Nutr Soil Sci 2012; 175: 171-179.

[29] Lehto T, Malkonen E. Effects of liming and B fertilization on B uptake of *Picea abies*. Plant Soil 1994; 163: 55-64.

[30] Fuertes ME, Lobartini JC, Orioli G.A. Boron nutrition, intracellular transport, and knife-cut disease in sunflower 2010; 41: 665-678.

[31] Pfeffer H, Danel F, Romheld V. Are there two mechanisms for boron uptake in sunflower? J Plant Physiol 1999; 155: 34-40.

[32] Dordas C, Chrispeels MJ, Brown PH. Permeability and channel-mediated transport of boric acid across membane vesicles isolated from squash roots. Plant Physiol 2000; 14: 1349-1361.

[33] Dordas C, Brown PH. Evidence of channel mediated transport of boric acid in squash (*Cucurbita pepo*). Plant Soil 2001; 235: 95-103.

[34] Liu GD, Wang RD, Liu LC, Wu LS, Jiang CC. Cellular boron allocation and pectin composition in two *Citrus* rootstock seedlings differing in boron-deficiency response. Plant Soil 2013; 370: 555-565.

[35] Huang L, Ye Z, Bell RW, Dell B. Boron nutrition and chilling tolerance of warm climate crop species. Ann Bot 2005; 96: 755-767.

[36] Pfeffer H, Dannel F, Romheld V. Boron compartmentation in roots of sunflower plants of different boron status: a study using the stable isotopes 10B and 11B adopting two independent approaches. Physiol Plant 2001; 113: 346-351.

[37] Dannel F, Pfeffer H, Romheld V. Update of boron in higher plants-uptake, primary translocation and compartmentation. Plant Biol 2002; 4: 193-204.

[38]   Brown PH, Hu H. Phloem B mobility in diverse plant species. Bot Acta 1998; 111: 331-335.

[39]   Mishra S, Hecakathom S, Frantz J, Futong Y, Gray J. Effects of B deficiency on geranium grown under different non-photoinhibitory light levels. J Am Soc Hortic Sci 2009; 134: 183-193.

[40]   Ishii T, Matsunaga T, Hayashi N. Formation of rhamnogalacturonan II: borate dimer in pectin determines cell wall thickness of pumpkin tissue. Plant Physiol 2001; 126: 1698-1705.

[41]   Dell B, Huang L. Physiological response of plants to low boron. Plant Soil 1997; 193: 103-120.

[42]   Lehto T. Effects of liming and boron fertilization on mycorrhizas of Picea abies. Plant Soil 1994; 163: 65-68.

[43]   Ruuhola T, Leppanen T, Julkunen-Tiito R, Rantala MJ, Lehto T. Boron fertilization enhances the induced defense of silver birch. J Chem Ecol 2011; 37: 460-471.

[44]   O' Sullivan J. Boron deficiency. Internet: http://keys.lucidcentral.org/keys/sweetpotato/key/Sweetpotato%20Diagnotes/Media/Html/TheProblems/MineralDeficiencies/BoronDeficiency/Boron%20deficiency.htm

[45]   Wojcik P, Treder W. Effect of drip boron fertigation on yield and fruit quality in a high-density apple orchard. J Plant Nutr 2006; 29: 2199-2213.

[46]   Davies BE. Applied Soil Trace Elements 1980; John Wiley and Sons, New York, p. 482.

[47]   Anonymous. Boron deficiency symptoms in the leaf of an oilseed rape or canola plant (*Brassica napus*). Internet: http://eu.art.com/products/p361002819-sa-i4013072/posters.htm?ui=56C42535DD1941FEAC753D859983F27D

[48]   Anonymous. Nutrient disorders of greenhouse Lebanese cucumbers. Internet: http://www.dpi.nsw.gov.au/agriculture/horticulture/greenhouse/pest-disease/general/cucumber-nutrition.

[49]   Dursun A, Turan M, Ekinci M, *et al.* Effects of boron fertilizer on tomato, pepper, and cucumber yields and chemical composition. Com Soil Sci Plant Anal 2010; 41: 1576-1593.

[50]   Esringu A, Turan M, Gunes A, Esitken A, Sambo P. Boron application improves on yield and chemical composition of strawberry. Acta Agriculturae Scand Section B-Soil and Plant Science 2011; 61: 245-252.

[51]   Antoniadis V, Chatzissavvidis C, Paparnakis A. Boron behavior in apple plants in acidic and limed soil. J Plant Nutr Soil Sci 2013; 176: 267-272.

[52]   Leite VM, Brown PH, Rosolem CA. Boron translocation in coffee trees. Plant Soil 2007; 290: 221-229.

[53]   Liakopoulos G, Psaroudi V, Stavrianakou S, Nikolopoulos D, Karabourniotis G. Acclimation of eggplant (Solanum melongena) to low boron supply. J Plant Nutr Soil Sci 2012; 175: 189-195.

[54]   An JC, Liu YZ, Yang CQ, Zhou GF, Wei QJ, Peng SA. Isolation and expression analysis of CiNIP5, a *citrus* boron transport gene involved in tolerance to boron deficiency. Sci Hortic 2012; 142: 149-154.

[55]   Byers DE, Mikkelsen RL, Cox FR. Greenhouse evaluation of four boron fertilizer materials. J Plant Nutr 2001; 24: 717-725.

CHAPTER 8

Molybdenum Deficiency

Abstract: The average concentration of Mo in the lithosphere is 2.4 mg/kg and in soils varies greatly, depending on parent material. Molybdenum solubility, thus its' availability for plants, is affected by many factors, such as soil pH, organic matter, the content of Fe oxides/hydroxides in soils, P concentration, liming, crop management, soil humidity and interactions with other nutrients. Molybdenum is mainly absorbed as MoO_4^{2-} and it is readily and highly mobile in xylem and phloem. The form by which Mo is translocated is probably that of MoO_4^-. From the physiological functions of Mo, the most important is that in atmospheric N capturing and uptake. Particularly, nitrogenase and nitrate reductase, which are involved in N fixation and NO_3^- reduction respectively, are Mo-containing enzymes and their activity is depressed under conditions of Mo deficiency. Apart from the involvement of Mo in nitrogen metabolism and fixation, Mo also participates in N transport in plants, as well as it occurs in more than 60 enzymes and catalyzes diverse oxidation and reduction reactions. Furthermore, Mo is also involved in processess concerning the synthesis of the phytohormones abscisic acid and indole-3 butyric acid.

Generally, Mo foliar level 0.5 p.p.m. is considered for most plant species as critical deficient, while for some others may be lower, for example 0.1 or 0.3 p.p.m. Molybdenum deficiency symptoms include deep chlorosis of old leaflets, spreading to young growth and intensification of chlorosis leading to bleaching. The leaf chlorosis of Mo deficiency somewhat resembles to that of N deficiency, due in part to Mo role in N utilization. In order to detect Mo deficiency before the appearance of symptoms, induciable nitrate reductase activity can be used as an indicator of the Mo nutritional status of plants, since Mo is an essential component of two major enzymes (nitrate reductase and nitrogenase). Finally, in order to correct Mo starvation, Mo can be supplied either in mixtures as fertilizers, or as seed coating, or foliar sprays of water-soluble Mo salts, mostly ammonium and sodium molybdates.

All the points referring to Mo availability in soils, its' uptake, the physiological roles of Mo in plant metabolism, the critical Mo concentrations, the symptoms of deficiency and the methods used for its' detection, as well as the Mo-fertilizers used to alleviate Mo stress in crops, are fully analyzed and discussed in this chapter.

Keywords: Mo adsorption, Mo availability, Mo deficiency, Mo fertilizers, Mo uptake, N capturing, N fixation, Nitrate reductase, Nitrogenase, NO_3 reduction.

INTRODUCTION

Molybdenum content in agricultural soils varies from about 0.2 to 36 mg/kg [1]; these values can widely vary, depending also on parent material. Soils derived from granitic rocks and argillaceous schists are often high in Mo, while highly weathered acid soils tend to be deficient [2]. The Mo content of soils usually resembles to that of their parent rocks and ranges from 0.013 to 17.0 p.p.m. [3]. The solubility and availability of Mo in soils is governed by many factors, such as pH, Fe oxides and hydroxides, drainage conditions, $CaCO_3$ content, salinity, organic matter *etc.* Particularly, in neutral and moderate alkaline pH soils the form that dominates is that of MoO_4^-; in contrast, at lower pH values $HMoO_4^-$ usually occurs [3]. Molybdenum is absorbed by plants as molybdate ion (MoO_4^-); this is why Mo deficiency for crops is most commonly observed in acid soils.

Two Mo-containing enzymes (nitrogenase and nitrate reductase) are involved in N metabolism, in N fixation and NO_3^- reduction, respectively, thus the requirement for Mo is greater for plants using NH_4-N, than for those using NO_3-N [3]. For that reason, inducible nitrate reductase activity can be used as an indicator of the Mo nutritional status of plants [2]. The critical Mo deficiency level in plants is very low (usually below 0.2 mg/kg d.w.). Generally, plants with a little less than 1 mg/kg d.w. are adequately supplied with Mo. Interveinal mottling, marginal chlorosis of the older leaves and upward curling of the leaf margins are typical symptoms of Mo deficiency. As the deficiency goes on, necrotic spots appear at leaf tips and margins, which are associated with high NO_3 concentrations [2]. In the cases of Mo deficiency, application of Mo salts is recommended. This supply can be done in soil, in a foliar form, or by seed treatment.

The purpose of this chapter is to collect and present all the data referring to: i) Mo availability in soils, ii) Mo uptake by plants, iii) the critical plant concentrations and symptoms of deficiency, iv) the physiological roles of Mo, as well as to v) Mo soil and foliar application rates.

Mo in Soils

Molybdenum is present in small amounts in the lithosphere (average 2.4 mg/kg) and in soils (ranging from 0.2 to 36 mg/kg) [1]. In aqueous solution with pH>4.3 Mo occurs mainly as MoO_4^- in its highest oxidized form (Mo(VI)). At lower pH (<4.3) protonated species ($HMoO_4^-$) become the prevailing forms [4]. Molybdenum content of soils depends greatly on parent material. Soils derived from granitic rocks and argillaceous schists are often high in Mo, while highly weathered acid soils tend to be deficient. Generally, 4 major fractions are present in soils: i) dissolved Mo, ii) Mo occluded with oxides (Al, Fe and Mn oxides), iii) Mo solid phases including molybdenite, powellite and ferrimolybdite, and iv) Mo associated with organic compounds [2].

Molybdenum is associated mainly with Fe oxides, probably as an adsorbed phase. The molybdate adsorbed on freshly precipitated $Fe(OH)_3$ is readily exchangeable, but as the precipitate ages Mo becomes less soluble and ferrimolybdite, or other slightly soluble Fe-Mo forms, may occur [3]. It has been known for many years that the adsorption of Mo in soils, like that for phosphate and sulphate, is strongly pH dependent, increasing with decreasing pH. This effect of pH on adsorption may be explained by competition between OH^- and MoO_4^{2-} ions for common adsorption sites. This pH dependence of Mo adsorption has practical importance, since Mo deficiency can be often controlled by liming [2].

Factors Influencing Mo Availability in Soils

pH, organic matter, crop management, soil texture and erosion, the quantity of Fe oxides and hydroxides in soil, P concentrations, interactions with other nutrients, parent material, soil acidification, liming, reduction conditions (E_h) *etc.* are among the most important factors influencing Mo solubility and availability for plants. The way by which each of these factors influences Mo availability is fully analyzed below.

I.   **pH.** The anion MoO_4^-, a form by which Mo is taken up by plants, dominates in neutral and moderate alkaline pH values. In contrast, $HMoO_4^-$ occurs at lower pH values. Generally, as pH falls, Mo soil concentration decreases. According to Therios (1996), Mo solubility is significantly decreased (due to reactions of Mo with Fe and Al and to the formation of insoluble subtances) in pH values lower than 5.5 [5]. Furthermore, according to Kabata and Pendias (2001), on acid soils (pH<5.5) low in Mo, and especially on those with a high Fe oxide level, Mo is hardly available to plants [3].

II.   **Organic matter.** Soils rich in organic matter can supply adequate amounts of Mo to plants due to slow release of this element from organic bound forms [3].

III. **Soil texture.** Generally, sandy soils are more prone to Mo deficiency, than the clayey ones, especially when they are also acid.

IV.   **Parent material.** In the USA the geographic pattern of Mo deficiency mainly follows the regions of acid sandy soils that are formed from parent material low in Mo [2].

V.   **The content of adsorbing oxides (*e.g.*, Fe oxides).** Adsorption of Mo on Fe oxides and hydroxides is a problem negatively influencing Mo availability. It is referred by Mengel and Kirkby (2001) that about 90% of the total Mo is considered unavailable for plants and in the mineral soil horizon 20-50% of Mo is present as amorphous and crystalline Fe oxides forms [2].

VI.   **P content.** The P-Mo interaction is often demonstrated as an enhancing effect of P on Mo availabiliy in acid soils. According to Bambara and Ndakidemi (2010), phosphate ions enhance molybdenum uptake by plants [6]. However, reported effects of P fertilizers on Mo availability are contradictory. It seems that P-Mo interactions are variable and highly governed by diverse soil factors [3].

VII. **Clay minerals.** Molybdate is adsorbed by clay minerals in an analogous way to phosphate [2].

VIII. **Crop management.** Crop management greatly influences soil fertility and productivity. The choice of only one plant species for cultivation has many times proved to be catastrophic for soil physical and chemical properties. In the case of Mo, the cultivation of soybean (a species having the ability for biological nitrogen fixation, so with enhanced activity of nitrogenase) over many years in Brazil has lead to decreased availability of Mo in soils and recently it is common to see that the crop generally responds positively to Mo fertilization [7, 8].

IX.   **Liming.** As Mo solubility increases with the increase of soil pH, liming seems to be a good management practice to increase Mo availability and uptake. According to Kaiser *et al.* (2005), the application of lime to agricultural soils has been an important tool to adjust soil pH and increase soluble molybdate [9]. Furthermore, it was found that Mo concentration in the shoots of soybean increased by a factor of 10 when pH increased from 5 to 7 by liming [4].

X.   **Soil humidity and wet conditions.** The solubility, and thus availability of Mo to plants, is highly governed by soil pH and drainage conditions. Molybdenum from wet alkaline soils is most easily taken up, but the geochemical processes involved in this phenomenon are not yet fully understood. In addition, soils in arid and semiarid regions, especially ferrasols, generally have higher Mo contents [3].

XI.   **Interactions with other nutrients.** Sulphate and molybdate are strongly competing anions during uptake by the roots. Therefore, sulphate containing soil amendments (such as gypsum), as well as single superphosphate, reduce Mo uptake [4].

Mo Uptake and Transport

Molybdenum is an essential micronutrient, but the physiological requirement for this element is relatively low. Plants take up Mo mainly as molybdate ions [3]. Although there is no direct evidence, there is a suggestion of the active uptake of Mo [10]. A molybdate-specific transporter has been identified in *Arabidopsis thaliana*. This transporter, MOT 1, has a high affinity for MoO_4^{2-} [11]. It is expressed both in roots and leaves and the protein appears to be localized in mitochondria [12]. In long-distance transport in plants, Mo is readily mobile in xylem and phloem [9]. Molybdenum is highly phloem-mobile and when applied as a foliar spray in the early stages is preferentially translocated into the nodules [13], while it is also very effective in increasing final yields. The form in which Mo is translocated is unknown, but its chemical properties indicate that it is most likely transported as MoO_4^{2-} rather than in complexed form [4].

Physiological Roles of Mo

Nitrogenase and nitrate reductase, which are involved in N fixation and NO_3^- reduction respectively, are Mo-containing enzymes and their activity (thus N metabolism) may be decreased by Mo deficiency. Particularly, nitrogenase, which is the key enzyme complex, unique to all N_2 fixing organisms, consists of two Fe proteins, one of which is the FeMo protein containing two unique metal centres, the P-cluster and the FeMo cofactor [14] (Fig. **1**). Legumes dependent on N-fixation have a high Mo requirement, particularly in root nodules. Nodule dry weight can increase 18-fold, which indirectly reflects the increase in the capacity for N_2 fixation by improved Mo supply [4]. Nitrate reductase is an enzyme with 3 electron-transferring prosthetic groups per subunit: flavin (FAD), heme and Mo cofactor (Moco) (Fig. **2**). Moco consists of Mo covalently bound to two S atoms in the tricyclic molecule pterin [4].

Due to the direct effect of Mo on NO_3^- reduction, plant growth is suppressed and less chlorophyll content is many times found. In the study of Datta *et al.* (2011) higher levels of chlorophyll a and b, as well as of total chlorophyll in the leaves of *Cicer arietinum* under Mo treatments, than under Mo starvation, have been found. These enhanced levels of chlorophyll may be ascribed-according to the same authors to a higher amount of N incorporated into the chlorophyll biosynthesis, since N is a constituent of this molecule [15]. Not only Mo deficiency, but also Mo excess may negatively influence the activities of nitrogenase and nitrate reductase. Significant reduction of nitrate reductase activity has been found at Mo concentration treatment of 6 p.p.m. in *Cicer arietinum* seedlings [15]. Apart from the involvment of Mo in nitrogen metabolism and fixation, Mo also participates in the transport of N in plants, as well as it occurs in more than 60 enzymes and catalyzes diverse oxidation and reduction reactions [16]. Some of the enzymes that require Mo for their activity, apart from nitrate reductase, are xanthine dehydrogenase, aldehyde oxidase and sulfite oxidase [9]. These enzymes play important roles in the response and resistance to various stresses [17].

According to Mengel and Kirkby (2001), Mo also plays another essential role in N metabolism and transport of some tropical and subtropical legumes, such as soybean and cowpea. In these legumes the dominant long distance N transport from root to shoot are the ureides, allantoin and allantoic acid [2]. More specifically, xanthine dehydrogenase is the responsible enzyme that plays a key-role in N metabolism [4]. Besides that role, xanthine dehydrogenase may also play a role in plant-pathogen interactions and cell death, associated with hypersensitive response and natural senescence [17]. Apart from nitrogen metabolism, Mo is involved in processess concerning the synthesis of the phytohormones abscisic acid and indole-3 butyric acid [9]. A key role in the first procedure (biosynthesis of ABA) plays the enzyme of aldehyde oxidase; it catalyses the conversion of abscisic aldehyde to abscisic acid (ABA), which is the last step in ABA biosynthesis [4]. Molybdenum applications to deficient wheat plants increased ABA concentration and cold tolerance [18]. Finally, Mo has a close relationship with metabolic products of C; it is referred that higher levels of Mo treatment may have promoted the content of C assimilation products, such as total soluble carbohydrates [15].

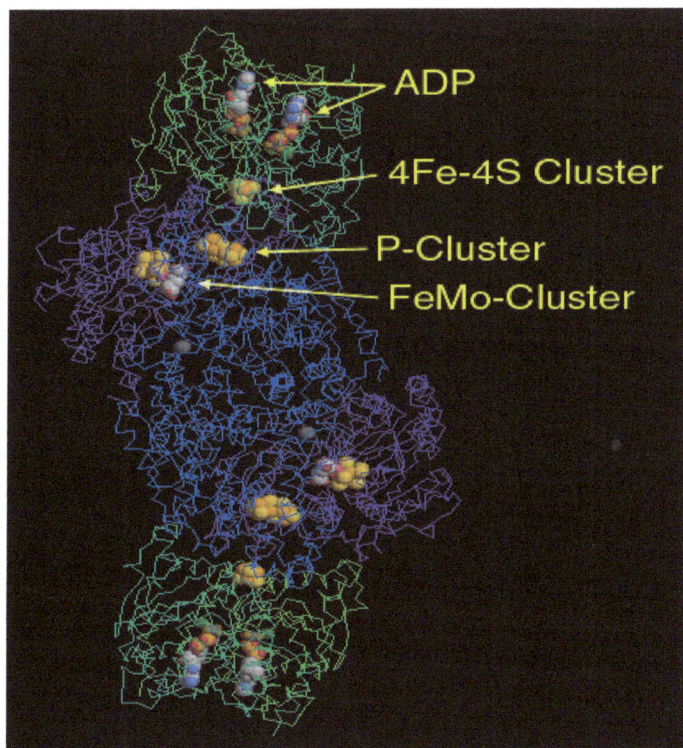

Fig. (1). The enzyme of nitrogenase (http://www.rcsb.org/pdb/101/motm.do?momID=26) [19].

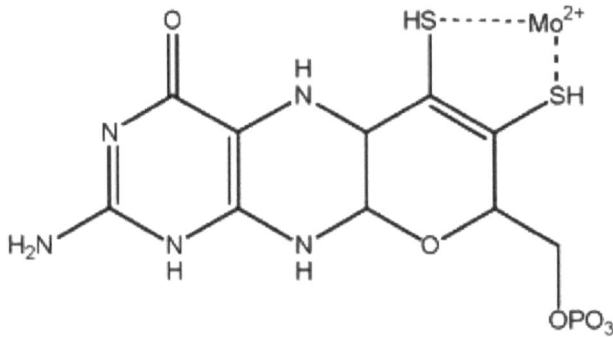

Fig. (2). The molecule of the enzyme nitrate reductase (http://faculty.virginia.edu/metals/cases/ east1.html) [20].

Mo Deficiency Symptoms and Critical Concentrations

Molybdenum deficiency symptoms include deep chlorosis of old leaflets, spreading to young growth and intensification of chlorosis, leading to bleaching [6] (Fig. **3**). Upward rolling of the leaves and leaf edge burn are also included within the most frequent Mo starvation symptoms. The leaf chlorosis of Mo deficiency somewhat resembles to that of N deficiency (Fig. **4**); in addition, stunned growth (due in part to the role of Mo in N utilization) occurs; however, these symptoms appear without the reddish coloration on the undersides of the leaves [21]. Distorted leaves can result under Mo deficient conditions to the failure of the interveinal areas to expand normally [22]. Affected by Mo deficiency leaves of *Cicer arietinum* were found to be dried and withered [23].

In alfalfa crop, symptoms of Mo deficiency are also like those of N deficiency, and appear as light green or yellow, stunted plants, caused by a lack of moly that is essential for N fixation by rhizobium bacteria that live on plant roots [24] (Fig. **5**). Generally, Mo deficiency is widespread in legumes and certain other plant species (*e.g.*, cauliflower and maize) grown in acid mineral soils, with large concentrations of reactive Fe oxidihydrate and thus a high capacity for adsorbing MoO_4^{2-} [4]. In maize plants, when Mo concentration is below 0.03 mg/kg in grains and 0.10 mg/kg in leaves, the risk of premature sprouting (Fig. **6**) increases [25]. In grapevine, Mo deficiency is associated with a symptom called Millerandage, which is characterized by unevenly developped grape bunches of

Fig. (3). Molybdenum deficiency symptoms on poinsettias [22].

berries, with varying size and degree of maturity [9] (Fig. **7**). The exact reason for this symptom is unclear, but it may be related to the effect of Mo on phytohormones [4]. Since Mo is an essential nutrient for nitrate reductase activity, the symptoms of Mo starvation are related to reduced reductase activity and yield. Indeed, spinach plants grown under deficient conditions of Mo revealed lower leaf nitrate reductase activity and yield with respect to the control (where adequate level of Mo was present) [15]. Molybdenum deficiency symptoms firstly appear on the upper leaves (Fig. **8**).

Molybdenum-deficient plants are more sensitive to low temperature stress [18] due to the effect of Mo on ABA biosynthesis. Moybdenum deficiency has also strong effects on pollen formation in maize; in Mo-deficient plants a large proportion of the flowers failed to open and the capacity of the anther for pollen production was reduced. Furthermore, the pollen grains were smaller, free of starch, while they had also lower invertase activity and showed poor germination [4].

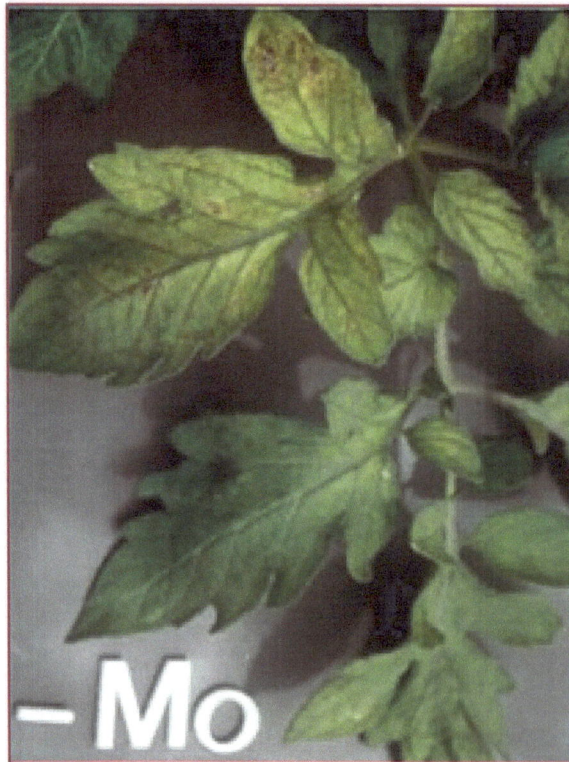

Fig. (4). Characteristic Mo deficiency in tomato plants [21].

Fig. (5). Contrast between Mo deficient and sufficient alfalfa plants [24].

Fig. (6). Premature sprouting in maize [26].

Fig. (7). The Millerandage symptom in grapevine [27].

Fig. (8). Molybdenum deficiency symptoms on the upper leaves of poinsettias [22].

Critical concentrations for Mo deficiency depend greatly on plant species. For example, for poinsettia, which has high Mo requirements, normal Mo concentrations vary from 1 to 5 p.p.m., while levels below 0.5 p.p.m. are considered deficient. For other greenhouse crops, the normal Mo level is lower, than for poinsettia [22]. In alfalfa crop it has been found that plant samples with 0.3 p.p.m. Mo were deficient, and samples containing from 0.4 to 1 p.p.m. Mo were marginal [24].

Detection and Assessment of Mo Deficiency

Plant tissue test seems to be the most frequently used way to confirm Mo deficiency. However, since Mo is an essential component of two major enzymes (nitrate reductase and nitrogenase), inducible nitrate reductase activity can be used as an indicator of the Mo nutritional status of plants [2]. Nitrate reductase activity is low in leaves of Mo-deficient plants, but it can be readily induced within a few hours by infiltrating the leaf segments with Mo [4]. An indirect method of assessing Mo deficiency is by determining the tissue nitrate-nitrogen (NO_3-N) level. Accumulation of NO_3^-N in levels >3,000 p.p.m. indicate Mo deficiency [22].

Mo Fertilization

Sources of Mo for crops are scarce, but Mo can be supplied either in mixtures as fertilizers, or as seed coating, or foliar sprays of water-soluble Mo salts, mostly ammonium and sodium molybdates (40% molybdenum) [28]. It is referred that a single application of 0.4 pounds per acre of molybdenum for alfalfa crop last from 5 to 15 years [24]. Moybdenum fertilization through foliar sprays can effectively supplement internal Mo deficiencies and rescue the activity of molybdoenzymes [9]. According to Williams *et al.* (2004), foliar sprays of Mo, applied before flowering, are effective in correcting Mo deficiency in grapevine [29]. Foliar sprays have a greater risk of plant injury. Early morning applications are preferred. Applications during the midday heat should be avoided. Plant uptake is enhanced by the increased drying time, which occurs during the moist conditions in the morning [22].

Another way of supplying Mo in crops is by enhancing the Mo content of seeds. According to Campo *et al.* (2009), only two foliar sprays of 400 g. Mo/ha each, with a minimum interval of 10 days between sprays, are capable of producing Mo-rich seeds of several soybean cultivars. As a result of this method, increases of as much as 3000% were obtained, in comparison to seeds obtained from plants, which did not receive any Mo application. Furthermore, in most cases Mo-rich soybean seeds did not require any further application of Mo fertilizer [30]. Finally, liming is also a very good strategy to increase Mo availability through the indirect effect of the increase of pH in acid soils. However, according to Marschner (2012), a combination of both liming and Mo supply often leads to luxury uptake and very high Mo concentrations in the vegetative parts of the shoots and seeds [4].

REFERENCES

[1]   Barber SA. Soil Nutrient Bioavailability. John Wiley and Sons, New York, 1984.
[2]   Mengel K, Kirkby E. Zinc. In: Mengel K, Kirkby E, Kosegarten H, Appel T (Eds) Principles of Plant Nutrition, 5th Edition 2001; Kluwer Academic Publishers, Dordrecht, The Netherlands, 2001.
[3]   Kabata-Pendias A, Pendias H. Trace Elements in Soils and Plants. 3rd ed. CRC Press, USA, 2001.
[4]   Marschner P. Mineral Nutrition of Higher Plants, 3rd ed. 2012.
[5]   Therios I. Mineral Nutrition of Higher Plants. Dedousis Publications. Thessaloniki, Greece, 1996. (In Greek).
[6]   Bambara S, Ndakidemi PA. The potential roles of lime and molybdenum on the growth, nitrogen fixation and assimilation of metabolites in nodulated legume: A special reference to *Phaseolus vulgaris* L. Afr J Biotech 2010; 8: 2482-2489.
[7]   Lantmann AF, Sfredo GJ, Borkert CM, Oliveira MCN. Resposta da soja a molibdenio em diferentes niveis de pH do solo. Rev Bras Cienc Solo 1989; 13: 45-49. (In Portuguese, English Abstact).
[8]   Campo RJ, Lantmann AF. Efeitos de micronutrientes na fixacao biologica do nitrogenio e produtividade da soja. Pesq Agropec Bras 1998; 33: 1245-1253. (In Portuguese, English Abstact).
[9]   Kaiser BN, Gridley KL, Brady JN, Phillips T, Tyerman SD. The role of molybdenum in agricultural plant production. Ann Bot 2005; 96: 745-754.
[10]   Moore DP. Mechanisms of micronutrient uptake by plants. In: 'Micronutrients in Agriculture', Mortvedt JJ, Giordano PM, Lindsay WL (Eds). Soil Sci Soc Am, Madison, WI, 1972:17.
[11]   Tomatsu H, Takano J, Takahashi H, Watanabe-Takahashi A, Shibagaki N, Fujiwara T. An *Arabidopsis thaliana* high-affinity molybdate transporter required for efficient uptake of molybdate from soil. Proc Nat Acad Sci USA 2007; 104: 18807-18812.

[12]   Baxter I, Muthukumar B, Park HC, *et al.* Variation in molybdenum content across broadly distributed populations of *Arabidopsis thaliana* is controlled by a mitochondrial molybdenum transporter (MOT1). Plos Genetics 2008; 4: 1-12.

[13]   Adams JF, Burmester CH, Mitchell CC. Long-term fertility treatments and molybdenum availability. Fert Res 1990; 21: 167-170.

[14]   Schwarz G, Mendel RR, Ribbe MW. Molybdenum cofactors, enzymes and pathways. Nature 2009; 460: 839-847.

[15]   Datta JK, Kundu A, Dilwar-Hossein S, Benerjee A, Mondal NK. Studies on the impact of micronutrient (molybdenum) on germination, seedling growth and physiology of Bengal Gram (*Cicer arietinum*) under laboratory condition. Asian J Crop Sci 2011; 3: 55-67.

[16]   Marschner H. Mineral Nutrition of Higher Plants, 2nd ed. Academic Press, London 1995; p. 75.

[17]   Schwarz G, Mendel RR. Molybdenum cofactor biosynthhesis and molybdenum enzymes. Ann Rev Plant Biol 2006; 57: 623-647.

[18]   Sun XC, Hu CX, Tan QL, Liu JS, Liu HG. Effects of molybdenum on expression of cold-responsive genes in abscisic (ABA)-dependent and (ABA)-independent pathways in winter wheat under low-temperature stress. Ann Bot 2009; 104: 345-356.

[19]   Goodsell D. Nitrogenase. Internet: http://www.rcsb.org/pdb/101/motm.do?momID=26

[20]   Anonymous. Metals in Medicine and the Environment. Problems with Molybdenum Deficiencies. Internet: http://faculty.virginia.edu/metals/cases/east1.html

[21]   Anonymous. Tomato crop guide: Nutrients deficiency symptoms. Internet: http://www.haifa-group.com/knowledge_center/crop_guides/tomato/plant_nutrition/nutrient_deficiency_symptoms/#{85431F47-75EF-4605-921A-C2690312B6E4

[22]   Anonymous. Poinsettia problem-Diagnostic Key-Corrective measures. Internet: http://www.ces.ncsu.edu/depts/hort/poinsettia/corrective/b14.html

[23]   Nautiyal N, Chatterjee C. Molybdenum stress-induced changes in growth and yield of chickpea. J Plant Nutr 2004; 173-181.

[24]   Freeman-Long R. Molybdenum deficiency in Alfalfa production. Internet: http://ucanr.edu/blogs/blogcore/postdetail.cfm?postnum=9735

[25]   Farwell AJ, Farina MPW, Channon P. Soil acidity effects on premature germination in immature maize grain. In: Baligar VC, Murrmann RP (Eds), Plant-Soil interactions at low pH, 1991; pp 355-361. Kluwer Academic Publishers, Dordrecht, The Netherlands; 1991.

[26]   Anonymous. Premature sprouting in maize (http://corn.osu.edu/newsletters/2012/2012-32/premature-sprouting-of-corn-kernels-thomison-and-geyer/image/image_view_fullscreen)

[27]   Anonymous. Vignobles Chain et fils. Glossaire Viti-Vinicole http://www.chaigne.fr/glossaire-viticole-vinicole-mot-98-Millerandage.htm

[28]   Mordvedt JJ. Sources and methods for molybdenum fertilization of crops. In: Gupta UC (Ed.). Molybdenum in Agriculture. Cambridge University Press, New York 1997; pp. 171-181.

[29]   Williams CMJ, Maier NA, Bartlett L. Effect of molybdenum foliar aprays on yield, berry size, seed formation, and petiolar nutrient composition of 'Merlot' grapevines. J Plant Nutr 2004; 27: 1891-1916.

[30]   Campo RJ, Araujo RS, Hungria M. Molybdenum-enriched soybean seeds enhance N accumulation, seed yield, and seed protein in Brazil. Field Crops Res 2009; 110: 219-224.

CHAPTER 9

Chlorine Deficiency

Abstract: Chlorine is an important micronutrient and despite the fact that plant tissues usually contain substantial amounts of Cl⁻, often in the range from 2 to 20 mg/g d.w., the demand for Cl⁻ for optimum growth is for most species considerably lower (deficiency symptoms usually occur in the range 70-700 μg/g d.w.). Chlorine is taken up by plants as Cl⁻ and it is highly mobile, so after absorption it can be easily transported inside plants. The negative charge of Cl in soil makes it prone to leaching in regions with high rainfalls. In contrast to that, in regions with high evapotranspiration (arid and semi-arid regions) Cl⁻ may be highly accumulated in surface soil horizons. Apart from the climatic conditions determining the accumulation or leaching of Cl⁻ in soils, the distance from the sea is another important factor influencing Cl concentrations in soils; so, Cl in soils exhibits a clear trend of decreasing concentration with increasing distance from the sea.

Chlorine is implicated in several physiological functions, such as in osmotic and stomatal regulation, in oxygen evolution in photosynthesis, in disease resistance and tolerance, as well as in fruit quality and crop yields. In recent publications it is referred that the critical Cl deficiency concentration is 2 g/kg d.w. (*i.e.*, 2000 mg/kg d.w.). Below that concentration Cl deficiency symptoms, such as chlorotic leaves, leaf spots, brown edges, restricted and highly branched root system, as well as wilting of leaves at margins and leaf mottling, may occur. In order to alleviate Cl starvation symptoms, some Cl-containing fertilizers that may be supplied to plants are those of KCl (47% Cl), $MgCl_2$ and $CaCl_2$ (64% Cl). Other (anthropogenic) sources of Cl supply to plants are the irrigation water, the use of de-icing salt to frozen roadways during winter months and the atmospheric pollution.

Keywords: Cl availability, Cl deficiency, Cl leaching, Cl uptake, osmoregulation, photosynthesis, stomatal regulation.

INTRODUCTION

The content of Cl⁻ greatly varies among soils. The plant available form of Cl is that of Cl⁻; this is why it tends to be leached from the surface layers to the deeper ones (mineral soils are negatively charged). Generally, Cl⁻ concentration in most well-developed soils is low, unless there is a good input from rainfall, fertilizers and irrigation [1]. In arid and semi-arid regions Cl⁻ tends to be accumulated in the surface horizons due to high evapotranspiration. Crops grown on these regions usually suffer from salt stress and Cl toxicity.

Chloride anion is readily taken up by plants and its mobility in short and long distance transport is high [2]. Chlorine in normal concentrations is an essential micronutrient of higher plants and participates in several physiological metabolism processes, such as in osmotic and stomatal regulation, in oxygen evolution in photosynthesis, and in disease resistance and tolerance [1]. In excessive concentrations Cl⁻ may be harmful to plants as a major component of salinity stress. Average Cl concentrations in plants are within the range from 2 to 20 mg/g. d.w., which is typical of the concentration farther of a macronutrient. In most plants species, however, the minimum Cl requirement for plant growth is within the range from 0.2 to 0.4 mg/g. d.w., *i.e.*, 10-100 times lower [2]. The main sources of Cl supply to plants are the irrigation water, rainwater, sea spray, fertilizers, dust and atmospheric pollution [3]. The quantities of Cl⁻ in the atmosphere and rain water are considerably influenced by the distance from the sea, falling off rapidly moving inland [4]. Due to the fact that there are many sources of Cl supply to plants, combined with the semi-arid climatic and soil conditions (high evapotranspiration, high salinity) prevailed in many regions of the world, more usual nutritional problem is that of Cl toxicity, than Cl deficiency.

Chlorine in Soils

Chlorine in soils is not adsorbed by minerals, and is one of the most mobile ions, being easily lost by leaching under freely drained conditions. Since it does not readily form complexes and shows little affinity in its adsorption to soil components, Cl⁻ movement within the soil is largely determined by water flows, and, in particular, the relationship between precipitation and evapotranspiration [3]. More specifically, chloride leaching must be taken into account; from the KCl applied in autumn, a substantial amount of chloride may be leached into deeper soil layers by winter rainfall, while K^+ remains adsorbed to soil colloids in the upper soil layer [4]. Soils high in Cl⁻ include those affected by the sea, or treated with irrigation water containing high Cl⁻ content and poorly drained soils receiving run off from other areas [4]. Chlorine in soils exhibits a clear trend of decreasing concentration with increasing distance from the sea [5]. It was found in Norwegian forest soils close to the sea a Cl range from 475 to 1806 p.p.m. Cl, with an average of 920 p.p.m., while in soils farther from the sea Cl was within the range from 174 to 375 p.p.m., with an average of 265 p.p.m. [6]. In addition, Cl in Japanese forest soils from the coastal plain ranged from 91 to 486 p.p.m., with an average of 228 p.p.m., while in topsoils from upland fields Cl ranged from 56 to 305 p.p.m., with an average of 114 p.p.m. [7]. In the study of Raji and Jimba (1999) total chlorine contents in 32 surface soils in Savanna zones of Nigeria varied from 47.2 to 296.5 mg/kg, while the water-soluble Cl varied from 0.1 to 7.7 mg/kg. Based on the above results, the authors supported that over 80% of the Nigerian savanna soils may be deficient in Cl [8].

The fraction of water soluble Cl⁻ in soils is a good indicator for Cl availability [4]. Anthropogenic activities, such as application of de-icing salt to frozen roadways during winter months, salt water spills associated with the extraction of oil, natural gas deposits and some coals [5] are included within the most important ones contributing to the enrichment of soils with Cl ions. Both water soluble Cl, as well as total Cl content were not significantly correlated with any of the basic soil properties (clay, organic matter, pH *etc.*) in 32 surface soils, studied in savanna zones of Nigeria. This lack of significant correlation between Cl and selected soil properties tends to indicate, according to Raji and Jimba (1999), that neither texture, nor organic matter and pH had a controlling influence on Cl occurrence in the study area [8].

Uptake and Transport

Chlorine is taken up by plants as Cl⁻ and it is easily transported inside the plants within the xylem or phloem. It is referred that Cl is relatively mobile within the phloem, and the circulation of Cl (defined as the ratio of phloem-xylem nutrient fluxes) is about 20% in a number of plants [3]. The uptake rate depends primarily on the concentration in the nutrient (or soil) solution and on the individual uptake potential of plant species. Chlorine uptake is metabolically controlled [4]. According to Kabata and Pendias (2001), the soluble Cl fraction in soils is taken up passively by roots and is easily transported in plants [5]. During the growth of a plant Cl⁻ is translocated from the root to the shoot *via* the xylem and is redistributed between tissues *via* the phloem. Chlorine can be also absorbed directly by plant leaves from aerial sources and some plants species are capable of extracting their total Cl requirements from aerial sources *via* their leaves [3]. Chloride uptake may be influenced by the presence of other anions in soil solution; so, it has been found that Cl⁻ uptake is inhibited by nitrate [4].

It has been recently identified a number of genes involved in Cl transport; chloride transporters in plants include members of the CLC protein family (chloride channels), which includes both Cl⁻ channels and (Cl⁻/NO_3^-) antiporters [9], since chloride uptake is inhibited by nitrate and *vice versa*, *i.e.*, nitrate uptake can be much depressed by high Cl⁻ concentrations [4]. Seven chloride channel (CLC) members have been identified in the *Arabidopsis* genome [10].

Roles and Physiological Functions of Cl in Plants

The most important roles of Cl in plants are those related with photosynthetic oxygen evolution, osmotic and stomatal regulation, as well as with disease resistance and fruit quality. All these physiological functions of Cl are explained in detail below.

Photosynthetic Oxygen Evolution

Chlorine is required for the functional assembly of the photosystem II cluster, comprising 4 Mn atoms. More specifically, Cl⁻ is required in the water splitting reaction in PSII; one or two Cl anions are required for the water oxidation cycle to proceed. The determination that only one or two Cl atoms are required for each water oxidation cycle within PSII suggests that the real Cl requirement for PSII function is lower than 1mM [2]. The presence of Cl⁻ was found to enhance both the evolution of O_2 and photophosphorylation. The absence of Cl⁻ in onion leaves inhibits stomatal opening [4].

Stomatal Regulation

The absence of Cl⁻ inhibits stomatal opening and impairs leaf water potential [4]. It has been recently found that opening and closure of stomata is mediated by fluxes of K and accompanying anions, such as malate and Cl⁻ [11]. In coconut, it was found that there is a close correlation between K and Cl fluxes during stomata opening from the subsidiary cells into the guard cells, and *vice versa* during stomata closure [2].

Osmoregulation

Chloride is a major active solute in the vacuoles and it is involved in osmoregulation [3]. Chloride has important functions in osmoregulation at different levels. At normal plant Cl concentrations it is a main osmoticum in the vacuoles of the bulk tissue (50-150 mM Cl), together with K. At low concentrations, which they are in the range of a micronutrient (about 1mM or below), these osmoregulatory functions of Cl are presumably confined to specialized tissues or cells, such as the extension zones of roots and shoots, as well as guard cells, where Cl concentrations may be substantially higher than the average of the bulk tissue [2]. It has been shown that Cl-treated plants can better regulate the moisture within their own cells and can also improve the management and uptake of moisture within the entire plant [12].

Disease Resistance and Tolerance

The suppression of root, stalk and leaf fungal diseases are aided by strategic chloride application. Research in Kansas, South Dakota, Texas, Oregon, Washington, New Jersey and Montana has demonstrated that diseases including take-all, root rot, Septoria, leaf rust, and tan spot (Fig. **1**), as well as glume blotch in small grains and stalk rot (fusarium) in corn often can be suppressed by chloride fertilization. Chloride aids in disease suppression by helping the plant develop a stronger cell wall, which in turn limits the damage the disease pathogen could cause when it is in contact with the plant. For example, the health of the flag leaf determines 70 to 80 percent of the yield potential in wheat. By protecting the flag leaf, chloride helps boost photosynthesis and enhances nutrient uptake, which helps ensure a higher potential yield [12].

Yields and Fruit Quality

Chloride improves the yields and quality of many crops, such as onions and cotton [1]. The studies on onion showed that it is a high Cl-requiring plant species and onions receiving Cl application throughout irrigation water had healthier, darker green foliage [13]. The increase in yields and fruit/crops quality due to Cl application may be probably ascribed to improved photosynthetic rates, as Cl participates in the photosynthetic oxygen evolution. For example, the lack of detectable diseases and spots on fruits, and especially on wheat, has been detected in several studies, clearly indicating that Cl was having a beneficial effect, which was more general than disease reduction [1, 14]. Engel *et al.* (1997) found that the origin of leaf spot occurrence in wheat plants was physiological and not infectious and it was highly linked to inadequate Cl mineral nutrition. More specifically, Cl fertilization (11-22 kg/ha) greatly suppressed or eliminated leaf spotting and increased yield in one or more cultivars at 6 or 7 sites [15]. In another experiment, it was found that leaf spot severity in wheat plants was closely related to shoot Cl concentration, and tissue necrosis was minor if Cl concentration was greater than 1g/kg d.w. In the same study, where it was tested if Br could play a similar beneficial role (as that of Cl) in suppressing tisue necrosis and improving crop yield of wheat, it was found that Br did not substitute for Cl by improving shoot and grain yield [16].

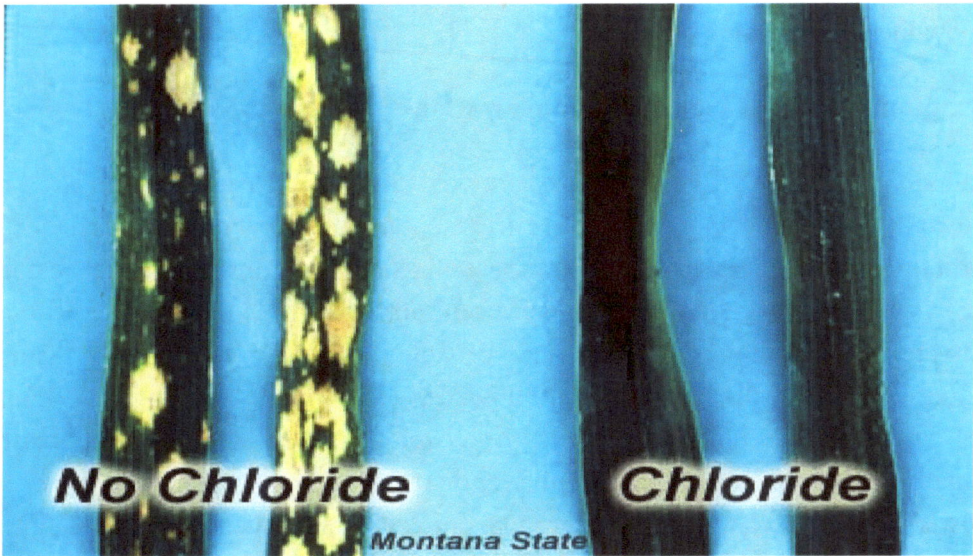

Fig. (1). On the wheat blades on the left, the occurrence of tanspot due to chloride deficiency is clearly visible [12].

Chlorine Deficiency and Critical Concentrations

Chlorine deficiencies are not observed often, but may occur when plants are growing in soils with Cl contents below 2 p.p.m. Generally, Cl is present in sufficient quantities in most soils to meet its requirements for most crops [5]. Plant tissues usually contain substantial amounts of Cl⁻, often in the range from 2 to 20 mg/g d.w. The demand for Cl⁻ for optimum growth, however, is for most species considerably lower, and deficiency symptoms occur in the range 70-700 μg/g d.w. [4]. According to Marschner (1995), a minimal Cl requirement for crop growth is 1000 mg/kg d.w. [17], while in more recent publications it is refered that the critical Cl deficiency concentration is 2 g/kg d.w. (*i.e.*, 2000 mg/kg d.w.) [2]. Plant species plays a major role in determining the critical deficiency concentration of Cl in the shoot dry weight and growth depression. For example, by withholding Cl supply growth was not affected in squash, but strongly reduced in lettuce. In addition, compared with most other plant species, kiwifruit has a very high Cl requirement. The reasons for the high Cl requirement of kiwifruit remain unclear [2]. According to Engel *et al.* (1997), the occurrence and severity of Cl-deficient leaf spot syndrome of wheat was cultivar-dependent; the same authors found that leaf spot damage increased exponentially as plant Cl dropped below the concentration of 1.0g/kg d.w. [15].

Symptoms of Cl Deficiency and Fertilization

Common symptoms of Cl deficiency are chlorotic leaves, leaf spots (Figs. **2**, **3**), brown edges and the appearance of restricted and highly branched root system. In addition, wilting of leaves at margins, as well as leaf mottling are typical features [18], while transpiration rates have been also found to be affected by Cl starvation. According to White and Broadley (2001), Cl deficiency causes reduced leaf growth and wilting, followed by chlorosis, bronzing and finally necrosis. Roots become stunted and the development of laterals is suppressed. Finally, fruits are decreased in number and size [3]. In sugar beet, leaf growth is slowed down owing to a lower rate of cell multiplication [4]. Chloride insufficiency in cabbage is marked by an absence of the cabbage odor from the plant [18]. In order to provide the necessary levels of Cl for crops in the case of starvation, the use of KCl fertilization is usually recommended. Indeed, much of the needed Cl in the Eastern U.S. Corn Belt region is replenished through the use of KCl (47% Cl) [12]. Other excellent sources of Cl include $MgCl_2$ and $CaCl_2$ (64% Cl) [18, 19].

Fig. (2). Physiological leaf spot under low Cl conditions. The symptoms are similar in appearance to tanspot or septoria [19].

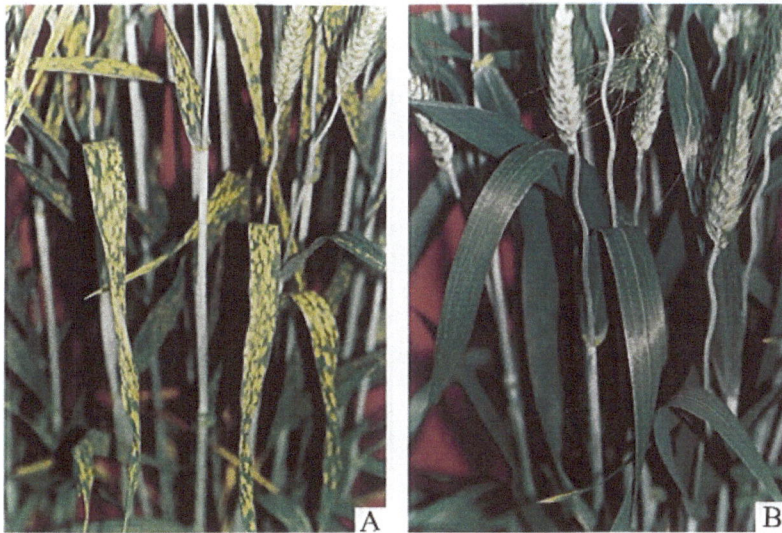

Chloride deficiency in Durum Wheat. A=No Cl, B=30 mmol Cl/pot

Fig. (3). Leaf spots in durum wheat plants due to Cl deficiency [18].

REFERENCES

[1]   Chen W, He ZL, Yang XE, Mishra S, Stoffella PJ. Chlorine nutrition of higher plants: Progress and Perspectives. J Plant Nutr 2010; 33: 943-952.

[2]   Marschner P. Mineral Nutrition of Higher Plants, 3rd ed. 2012.

[3]   White PJ, Broadley MR. Chloride in soils and its uptake and movement within the plant: A review. Ann Bot 2001; 88: 967-988.

[4]   Mengel K, Kirkby E. Chlorine. In: Mengel K, Kirkby E, Kosegarten H, Appel T (Eds) Principles of Plant Nutrition, 5th ed. 2001; Kluwer Academic Publishers, Dordrecht, The Netherlands 2001; pp. 639-643.

[5]   Kabata A, Pendias H. Trace elements in soils and plants. 3rd ed. CRC Press, USA, 2001; pp. 279-280.

[6]   Lag L, Steinnes E. Regional distribution of halogens in Norwegian forest soils. Geoderma 1976; 16: 317.

[7]   Yuita K. Iodine, bromine and chlorine contents in soils and plants of Japan. Soil Sci Plant Nutr 1983; 29: 403.

[8]   Raji BA, Jimba BW. A preliminary chlorine survey of the savanna soils of Nigeria. Nutr Cycl Agroecosys 1999; 55: 29-34.

[9]   Zifarelli G, Pusch M. CLC transport proteins in plants. FEBS Lett 2010; 584: 2122-2127.

[10]   Isayenkov S, Isner JC, Maathuis FJM. Vacuolar ion channels: roles in plant nutrition and signaling. FEBS Lett 2010; 584: 1982-1988.

[11]   Roelfsema MRG, Hedrich R. In the light of stomata opening: new insights into the 'Watergate'. New Phytol 2005; 167: 665-691.

[12]   EvansEnterprises. Essential nutrient-Chloride. Internet: http://www.evansenterprises.net/E E-EssentialNutrients.shtml

[13]   Randle WM. Chloride requirements in onion: Clarifying a widespread misunderstanding. Better Crops 2004; 88: 10-11.

[14]   Fixen PE, Buchenau GW, Geldermann RH, *et al*. Influence of soil and applied chloride on several wheat parameters. Agron J 1986; 78: 736-740.

[15]   Engel RE, Bruckner PL, Mathre DE, Brumfield SKZ. A chloride-deficient leaf spot syndrome of wheat. Soil Sci Soc Am J 1997; 61: 176-184.

[16]   Engel RE, Bruebaker L, Emborg TJ. A chloride deficient leaf spot of durum wheat. Soil Sci Soc Am J 2001; 65: 1448-1454.

[17]   Marschner H. Mineral Nutrition of Higher Plants. 2nd ed. Academic Press, London, 1995.

[18]   Agronomic library. Chloride. Internet: (http://www.spectrumanalytic.com/support/library/ ff/Cl_Basics.htm).

[19]   Mikkelsen R. Managing plant nutrients. Take another look at chloride? Internet: http://managingnutrients.blogspot.gr/2012/11/take-another-look-at-chloride.html

INDEX

A

ABA biosynthesis   163

Ability, differential Mn uptake   98

Abnormalities, floral   139, 140

Acid

    citric   46, 47

    organic   40, 41, 44, 49, 53

    oxaloxic   22

    strong   123

Acid fertilizers   48

Acid mineral soils   164

Acid phosphatases   22, 34, 88, 89, 90, 100

Acid sandy soils   82, 160

Acid soils   6, 12, 105, 114, 126, 151, 158-160, 170

    limed   151

    weathered   158, 159

Adsorption, boron   142, 143

Adsorption of Zn^{+2} 65, 67

Aerial plant parts   144

Alkaline   5, 6, 80, 111, 114, 125, 131

Alkaline soil conditions   38, 39, 56, 94, 142

Alkaline soils   4, 6, 39, 57, 67, 68, 82, 109, 114, 126, 139, 142

    wet   161

Alleviation   63, 65

Alluvial soils   105

Amaranthus retroflexus   57, 62

Aminoacids   100, 101

Amphibolites   91, 92

Amylohydrolase, allantoate   22, 102

Analysis, leaf nutrient   20

Andesite   90-92

Anions   142, 161, 175

F

I

K

L

M

P

T

U

www.ingramcontent.com/pod-product-compliance
Lightning Source LLC
Chambersburg PA
CBHW050845220326
41598CB00006B/439